Stars
and Planets

Stars
and Planets
Understanding the Universe

Giles Sparrow

Revised edition published in 2015

© 2008 by Amber Books Ltd.

First published in 2008

Reprinted in 2016

PICTURE CREDITS
All images courtesy of NASA except for the following: Art-Tech/IMP: 8, 10; Dorling Kindersley: 7, 12; David Jewitt: 96; Science Photo Library: 46, 79, 202(Eckhard Slawik), 203(John Sanford), 215(Mount Stromlo and Siding Spring Universities), 218(Rev. Ronald Royer), 219-220(John Sanford), 222(Chris Butler), 224(John Chumack), 225(Robin Scagell), 227(John Sanford), 231(Eckhard Slawik), 236(Eckhard Slawik), 238(Eckhard Slawik), 240(M. Karovska, Harvard CFA), 241(Pekka Parviainen), 242(Eckhard Slawik), 243(John Sanford), 244(Mark Garlick), 246(Davide de Martin), 261(David A. Hardy/PPARC), 263(European Southern Observatory), 272(R Ibata, Strasbourg Observatory), 279(National Optical Astronomy Observatories), 287(NOAO/AURA/NSF), 296(NOAO/AURA/NSF); Fahad Sulehria/Novacelestia.com: 226, 228, 247, 257, 260

Published by
Amber Books Ltd
74–77 White Lion Street
London
N1 9PF
United Kingdom
www.amberbooks.co.uk
Appstore: itunes.com/apps/amberbooksltd
Facebook: www.facebook.com/amberbooks
Twitter: @amberbooks

ISBN: 978-1-78274-260-9

Project Editor: Sarah Uttridge
Design: Andrew at Ummagumma

Printed in China

CONTENTS

Introduction

The night sky has been a source of endless fascination for people since before the beginning of recorded history. Countless ancient monuments aligned to the movements of the Sun, Moon and stars, scattered around the world and often displaying staggering levels of sophistication, bear witness to the significance our ancestors attached to the heavens.

Since those ancient times, our understanding of the Universe around us has changed immeasurably, transforming the points of light in the sky into alien worlds and distant suns. We believe we understand the histories of the planets in our Solar System, the billion-year life cycles of the stars themselves, and even the place of our galaxy in the unimaginably vast Universe. At the same time, technology has evolved to a point where man has walked on the Moon, robot probes have investigated all the major planets, and telescopes orbit hundreds of kilometres above our heads, monitoring the sky in radiations that are far beyond the range of the mere human eye.

And yet still the fascination remains, perhaps because the night sky connects us directly with the awe and wonder of the cosmos. Understandably, though, many people find astronomy somewhat daunting – there are countless exotic types of object, bizarre and bewildering names, and counterintuitive concepts that can be hard to grasp – even the stars in the sky seem too many to count. Hopefully this book will convince you otherwise.

EARTH IN SPACE

The Earth is our platform for observing the Universe, so it's important to understand something of how its motions through space affect our point of view. Most noticeable of all, the Earth spins on its axis once every 23 hours 56 minutes, so that for roughly half the time, the Sun is above the horizon, and its glare, coupled with the brightening of the atmosphere created by its scattered light, drowns out the comparatively faint light of more distant objects. When the Sun sets and the stars emerge, the sky appears to spin around a fixed point in the sky, a 'celestial pole' that is lined up directly with Earth's axis of rotation. The precise position of this celestial pole depends on an observer's latitude on Earth – as they get closer to the equator, it sinks lower in the sky towards the pole of their hemisphere, and the circle of 'circumpolar' stars (those close enough to the pole to never set) gets smaller.

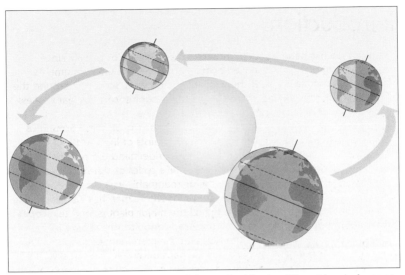

Earth's orbit affects the amount of sunlight received by the different hemispheres at different times of year.

As the Earth rotates, stars set in the west and rise in the east throughout the night – only the circumpolar stars, and those directly opposite the Sun, remain in view throughout the night.

As the Earth rotates, it is also orbiting the Sun once every 365.25 days. This means that from our point of view, the Sun is slowly moving eastwards around the sky, and it takes a little more than a full rotation of the Earth for the Sun to return to the same direction in the sky. The average period is more or less exactly 24 hours. Furthermore, because the Earth is tilted at an angle of 23.5°, the amount of sunlight received by each hemisphere varies throughout the year, creating the seasons. If we could halt the Earth's rotation and look at the Sun's movement against the 'fixed' background stars, we would see that it follows a wavelike path through the sky during the course of the year, appearing against different stars at different times. This path is known as the ecliptic, and it represents the plane of Earth's orbit around the Sun, projected into the heavens. Indeed, despite the fact that science long outgrew it, astronomers still make use of the old idea that the

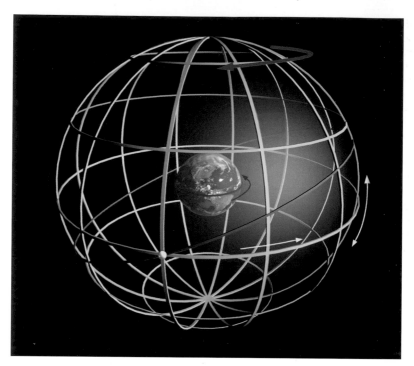

The celestial sphere projected from Earth's poles and equator appears to rotate around the Earth once a day. The purple line shows the plane of the ecliptic, the yellow dot the first point of Aries, and the arrows indicate directions of right ascension and declination.

entire sky is a spherical shell wrapped around the Earth. This 'celestial sphere' forms the basis for the system of 'celestial co-ordinates' used to locate objects in the sky. While places on Earth can be located by their latitude and longitude relative to the equator and the Greenwich Meridian, points in the sky can be pinned down by analogous co-ordinates called declination (DEC)and right ascension (RA). In this case, declination is measured in degrees, minutes and seconds of arc relative to the celestial equator – a projection of Earth's own equator onto the sky. Objects on the pole have 0°

declination, while the north celestial pole has a declination of +90° and the south celestial pole -90°. Right ascension is measured relative to the 'First Point of Aries' – the point where the Sun's track along the ecliptic passes from the southern to the northern hemisphere of the sky across the celestial equator at the northern spring equinox. Somewhat confusingly, though, right ascension is not measured in degrees, but in hours, minutes and seconds of time: an object's RA indicates the time it takes to 'follow' the First Point across an observer's meridian (the north-south line across their sky), measured anticlockwise around the celestial sphere.

Although this system of 'equatorial co-ordinates' is sufficient to pinpoint any object in the sky, it is not terribly intuitive, and as a result astronomers still make use of the patterns of bright stars first devised by their prehistoric ancestors. There are 88 of these constellations in total, ranging in size from the enormous but faint Hydra (the Water Snake) to the tiny but brilliant Crux (the Southern Cross). The patterns have no real significance – typically they link stars that are far apart in space – but they are a useful way of splitting up the sky. Since 1930, the constellations have actually been defined as specific areas of sky with definite boundaries, so that any object can be allotted to one constellation or another based on its celestial co-ordinates.

As the Sun crawls around the sky along the ecliptic, it passes through a dozen of the most ancient and well-known constellations – the signs of the zodiac. In fact, the Sun passes through thirteen constellations in all – the extra one, Ophiuchus, is sometimes known as the 'thirteenth sign'. It has shouldered its way into the zodiac because the stars have gradually changed their orientation in the sky over the past few thousand years. This steady drift, known as precession, is caused by the long, slow wobble of Earth's axis of rotation, which rotates like the axis of a spinning top, once every 25,800 years. Fortunately, it is small enough to have little effect from year to year, and star atlases only have to be recompiled, and co-ordinates updated, every 50 years.

THE SOLAR SYSTEM

The Earth, of course, is not the only object moving around the Sun. It is the third of eight major planets of varying sizes, which orbit among myriad smaller objects in the gravitational thrall of our star.

Closest to the Sun are four relatively small, rocky planets (Mercury, Venus, Earth and Mars), of which Earth is the largest. Further out, beyond a rocky belt of smaller worlds called the Asteroid Belt, lie four much larger giant planets composed largely of gas and liquid. And beyond these lies a doughnut-shaped ring of smaller icy worlds, the Kuiper Belt. Pluto, accepted for a long time as a ninth major planet, is in fact just the brightest of these distant dwarfs. At the very edge of the Sun's gravitational reach, the entire Solar System encased in a spherical shell of trillions of comets called the Oort Cloud.

Between and beyond these worlds orbit countless chunks of rocky and icy debris – rocky meteoroids and asteroids, and icy comets and centaurs. Most of the planets are also accompanied by moons – natural satellites that either formed with them when the entire Solar System collapsed out of an enormous gas cloud 4.6 billion years ago, or were captured into orbit at a later stage.

Aside from the Sun, Earth's Moon is the most obvious celestial object, and certainly the easiest to observe. It lies an average of 400,000km (248,548 miles) away, and even to the naked eye it clearly shows a range of surface features, including craters, lava plains and mountain ranges. The Moon orbits the Earth every 27.3 days, and as it does so, it displays different amounts of its sunlit face to Earth, resulting in a cycle of phases that goes from New Moon, when the sunlit side is completely hidden, through crescent and 'first quarter' (Half Moon), to bloated gibbous and then full, before diminishing again. Because the Moon's position in the sky has to 'catch up'

The moon's phases change depending on whether it lies near the Sun, or on the opposite side of the sky from it.

with the Sun's in order to return to New Moon, the cycle of phases takes 29.5 days on average – slightly longer than a lunar orbit.

The vast majority of worlds in the Solar System orbit in roughly the same flat plane as the Earth, and so they usually stay close to the ecliptic as they make their own way around the Sun. The shape of a planet's path around Earth's skies depends on the relative positions of it, the Sun and the Earth: for example, the 'inferior' planets Mercury and Venus, which orbit closer to the Sun, never move far away from it in the sky. Mercury only ever makes fleeting appearances in the twilight, looping into the sky for a few days, either shortly before sunrise or just after sunset. Venus has a longer, slower orbit but it too is anchored to the Sun and eventually pulled back. As the brightest object in the sky after the Sun and Moon, Venus also displays Moonlike phases that are easily seen through small telescopes: when it lies on the near side of the Sun and close to Earth, it appears as a thin crescent; when it is furthest from the Sun it appears as first- or last-quarter, and when it is on the opposite side of the Sun it appears almost 'full'. The phases have the same cause as those seen on the Moon – the changing amount of the planet's sunlit half that is visible from Earth.

In contrast to the inferior planets, the outer 'superior' planets are not anchored to the Sun, so they circle the entire sky in periods that are roughly equivalent to their orbits. They show no phases because we only ever see their sunlit sides, but they do vary in brightness and apparent size between 'conjunction', when they lie on the opposite side of the Sun from Earth, and 'opposition', when they are on the same side of the Sun as Earth, and at their closest to us.

THE STARS

So far away that even their light takes years to reach us, the stars form a 'fixed' backdrop to the faster-moving worlds of the Solar System. However, each star is a sun in its own right – an enormous ball of exploding gas, potentially at the centre of its own solar system. In between the stars lie vast clouds of gas and dust – nebulae that mark the beginnings and endings of stellar life cycles.

Even a glance at the night sky will show that the stars have different properties – they vary wildly in brightness and colour. The apparent brightness of a star as seen from Earth is measured using a system called 'magnitude', in which the brighter a star is, the lower its magnitude.

The faintest naked-eye stars have magnitudes around 6.0, while the brightest stars of all are so brilliant they have negative magnitudes – Sirius's magnitude is -1.46. Binoculars, which have larger lenses to gather in light from faint objects, can extend our eyesight down to around magnitude 9, while a small telescope can take it down to about magnitude 12.

Because stars can lie at wildly varying distances, the apparent magnitude of a star seen from Earth does not necessarily represent its true brightness. Fortunately, there are various ways to estimate the distance to stars and reveal their true brightness – measured here in terms of luminosity compared to the Sun. Meanwhile, the colours of stars can also reveal many of their true properties, such as their surface temperatures (just as a white-hot iron bar is hotter than a red-hot one, so a white star is hotter than a red one). Because the surface temperature of a star depends on the amount of surface through which its energy can escape, astronomers can work out the size of a star from its luminosity and colour – for instance, if a hot blue star and a cool red one have similar luminosities, then the blue one must be significantly smaller.

This map of the Solar System indicates the positions of the major planets, the asteroid belt and the dwarf planet Pluto.

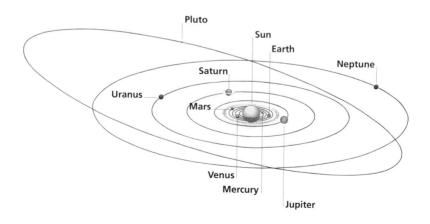

Since stars live and die on timescales of millions or billions of years, we rarely get to see major changes taking place. However, by measuring the properties of present-day stars and using the laws of statistics, along with our knowledge of the nuclear processes that take place inside stars, we can piece together a picture of their typical life stories.

Around 1910, two astronomers had the idea of plotting a chart to compare the luminosities and surface temperatures of stars. The result was the Hertzsprung-Russell diagram, which reveals that the vast majority of stars lie along a band called the 'main sequence', ranging from faint red stars to brilliant blue ones. Every star except for the most massive spends the majority of its life in one place on the main sequence, its position determined by its mass – the more massive a star is, the brighter it will shine for the majority of its life, and the hotter its surface will be. The penalty for this brilliance, though, is that it dooms the star to a much shorter life than more sedate, lower-mass stars like our Sun.

When a star nears the end of its life and begins to run short of fuel, it brightens considerably, but swells and cools, in its attempts to keep shining – transforming into an orange or red giant. Eventually, with its fuel exhausted, it dies, either casting off its shells in a beautiful planetary nebula or (if it has enough mass) detonating in a supernova explosion. The burnt-out remnants left behind by such events are some of the strangest objects in the Universe.

OUR GALAXY AND OTHERS

All the stars in the sky are members of our own spiral galaxy, the Milky Way, but ours is just one of perhaps a hundred billion galaxies in the wider Universe. These range in size and shape from tiny, shapeless and sparsely populated dwarfs to enormous, sharply defined spirals and even bigger elliptical balls of stars. Compared to their size, galaxies are far more closely packed than stars – they gather together in clusters that merge at their edges to form larger superclusters. On the largest scales of all, the superclusters form the knots and strings in a weblike mesh of filaments surrounding vast and apparently empty voids in the Universe. Beyond the immediate neighbourhood of the Milky Way, galaxies are among the most challenging objects to observe, and certainly require a good telescope, but they give a true picture of the immensity of the cosmos and the infinite depth of its appeal. Even after millennia of discovery, the Universe is still rich in mystery and filled with beauty and wonder.

The Sun

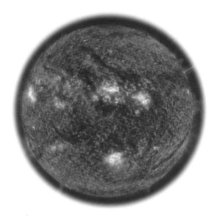

The Sun is our local star, a ball of exploding gas at the centre of our solar system, whose gravity, heat and light dominate a region of space about two light years in diameter. An average star – a little on the small side – it is composed almost entirely of the light gases hydrogen and helium. It shines because of nuclear fusion reactions at its core, which forge atoms of hydrogen, the simplest and lightest element, into helium, the next simplest, releasing energy in the process. The visible surface of the Sun, known as the photosphere, does not mark its true outer edge – it is merely the region where its layers of gas become transparent. The sparse outer layers of the Sun, its corona or atmosphere, stretch for millions of kilometres further into space, blending into the solar wind of particles blown out across the solar system.

Type of object:	Star
Average distance from Earth:	149.6 million km (92.9 million miles)
Equatorial diameter:	1.4 million km (864,900 miles)
Rotation period:	25–34 days
Mass:	333,000 Earths
Luminosity:	385 million billion billion watts

Interior of the Sun

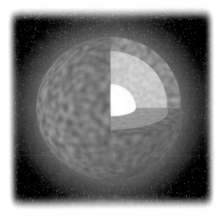

The Sun's interior consists of three layers. At the core, with a diameter one-quarter of the entire Sun, temperatures and pressures are so high that nuclear fusion reactions take place. Hydrogen nuclei (the cores of hydrogen atoms) are forced together to build the nuclei of helium; it takes four hydrogen atoms to create one helium atom, releasing energy. In the radiative zone, stretching from above the core to about two-thirds of the Sun's diameter, radiation from the centre works its way out through the Sun, bouncing between densely packed atomic nuclei and losing energy. At the base of the outermost convective zone, the Sun's gases become opaque. They absorb the radiation from below and heat up, rising to the surface to release their energy at the photosphere, where the Sun becomes transparent again.

Core density:	150 tonnes/cubic metre
Core temperature:	13.6 million °C (24.5 million °F)
Radiative zone:	Density: 10 tonnes/cubic metre
	Temperature: 5 million °C (9 million °F)
Convective zone:	Diameter: 1.4 million km (864,900 miles)
	Density: 200 kg/cubic metre at base
	Temperature: 2 million °C (3.6 million °F) at base.

Photosphere

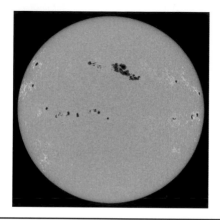

The Sun's visible surface is a layer 100km (62 miles) deep where the Sun's gases become transparent to visible light. The photosphere's distinctive yellow-white colour is due to its temperature (5500°C/9900°F), at which the Sun emits radiation with a broad spectrum of colours that combine to form yellowish light. The blazing surface is too bright to look at directly, but filters reveal a seething granular appearance created by the tops of convection cells rising from below. Dark sunspots on the photosphere are regions where the temperature is much lower – 3500°C (6300°F). Sunspots form in areas where the Sun's magnetic field pushes out through the photosphere in loops. These vary in number and location throughout a 22-year cycle that is driven by changes to the strength and shape of the solar magnetic field.

Depth:	Approx. 400km (240 miles)
Density:	0.2g/cubic metre
Temperature:	7340°C (13,240°F)
	4200°C (7590°F)
Typical features:	Sunspots, faculae (bright streaks), granules, supergranules

Chromosphere

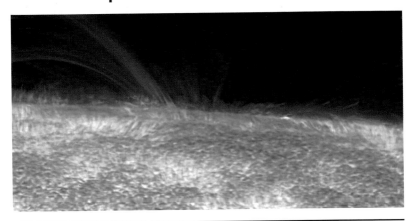

The transparent region of the Sun above the photosphere is called the chromosphere (from the Greek for 'colour') because of its red tint. Here, opaque features rising out of the photosphere can be seen – around the edge of the Sun during solar eclipses; at other times, specially tuned telescopes reveal them. The most prominent are spicules – flamelike pillars that rise and sink in minutes, and which may carry energy from the photosphere into the 'transition region' at the base of the corona. Loops of the Sun's magnetic field channel rivers of cooler dense gas through it. Silhouetted against the photosphere, they form sinuous 'filaments', but when the Moon blocks our view of the photosphere during a total solar eclipse, the loops stand out around the edge of the Sun as red or pink 'prominences'.

Depth:	2000km (1200 miles)
Density:	About 1 millionth of a gram/cubic metre
Temperature:	4500°C (8100°F) at base 20,000°C (36,000°F) at outer surface
Typical features:	Spicules, filaments, prominences, plages (bright patches around sunspots)

Corona

The outer layers of the Sun's atmosphere are sparse but hot – far hotter than the photosphere and chromosphere below. Astronomers are not sure how the Sun heats its corona to temperatures of 1–2 million °C (2–4 million °F), but suspect that some of the energy comes from shock waves originating deep inside the Sun, and some from the solar magnetic field. Throughout the solar cycle, the corona varies in shape and density. Sometimes it is compact and uniform, but when the Sun is at its most active it becomes patchy and larger, fed by solar flares and 'coronal mass ejections' (CMEs). Flares and CMEs are triggered when loops in the Sun's magnetic field 'short-circuit' and reconnect closer to the photosphere. The excess energy released ejects plumes of gas from the photosphere, at close to the speed of light.

Diameter:	Variable with solar activity (approx. 1 million km)
Density:	About 0.0002 billionths of a gram/cubic metre
Transition region temperature:	20,000°C (36,000°F) at base 1 million °C (1.8 million °F) at outer surface
Outer corona temperature:	2 million °C (3.6 million °F)
Typical features:	Coronal loops and holes, solar flares, coronal mass ejections

Solar wind

Particles are continually blown from the surface of the Sun and across the solar system, forming the solar wind. Travelling at hundreds of kilometers a second, the solar wind particles carry with them the Sun's magnetic field, and form spiral 'sheets' of differing magnetic polarity that stretch to the edge of the solar system. The strength of the wind, and the shape of its currents, is governed by the solar cycle. As the solar wind passes the planets, particles are caught up in the planetary magnetic fields and ricochet back and forth, gaining energy until being channelled down on to the planet's magnetic poles. On planets with substantial atmospheres such as Earth, collisions between the particles and air molecules produce radiation in various colours, forming aurorae like the aurora borealis and aurora australis.

Extent:	Termination shock (drop to subsonic speed) around 85 AU
Solar mass loss to solar wind:	1 million tonnes/second
Speed near Earth:	Around 450km/s (280 miles per second)
Composition:	Dominated by ionized hydrogen (proton and electron particles), 8 per cent helium ions
Trace heavy ions:	Carbon, nitrogen, oxygen, neon, magnesium, silicon, sulphur, iron

Mercury

The smallest of the major planets, Mercury is the closest to the Sun, and as a result it whizzes around its orbit at a speed of 48km (30 miles) per second. This tiny rocky world has a huge iron core that is out of proportion with its size, and most astronomers think this is evidence of a collision early in Mercury's history that blasted away much of the material from its outer layers, and reformed the planet. Today Mercury is a dead world, cratered like Earth's Moon. Its rotation has slowed, so it spins on its axis in precisely two-thirds of its year, and a curious result is that sunrises on the planet are separated by two Mercury years. Temperatures on the daylit side of the planet reach 430°C (810°F) – but it seems that ice, perhaps dumped during comet collisions, survives in permanently shadowed craters near the poles.

Type of object:	Rocky planet
Closest to Sun:	46 million km (28.6 million miles)
Furthest from Sun:	69.8 million km (43.4 million miles)
Orbital period:	88 days
Diameter:	9758.8km (6061.4 miles)
Rotation period:	58.6 days
Axial tilt:	0.01°

Caloris Basin

The most impressive feature on Mercury is this impact basin, 1300km (808 miles) across. It formed four billion years ago, during the 'late heavy bombardment', when the larger worlds mopped up the dozens of rogue planetoids that survived in the inner solar system. The only images of it come from *Mariner 10*, which flew past Mercury three times in 1974 and 1975, but photographed only half the planet (and half the basin). They reveal a flat central plain, 2km (1.2 miles) below its surroundings, ringed by concentric mountain ranges and radiating ridges formed by the 'ejecta' debris flung away from the impact. Like the lunar seas, the lowest parts of this gouge in Mercury's surface seem to have been flooded by lava from inside the planet – but without the obvious colour difference seen on our Moon.

Feature of:	Mercury
Type:	Impact basin
Diameter:	1350km (840 miles)
Height of surrounding mountains:	2km (1.2 miles)
Age:	4 billion years

Weird Terrain

One of the most bizarre landscapes photographed by *Mariner 10* during its series of Mercury flybys was the 'Weird Terrain', a jumbled mixture of blocks, deep furrows and hummocky hills that looks as though the entire landscape has been shaken up and then left to settle back in a new configuration. In fact, this seems to be exactly what did happen; the clue lies in the Weird Terrain's location on Mercury – precisely opposite the heart of the Caloris Basin. It seems that shockwaves from the Caloris impact were felt all the way around the planet. Some reverberated straight through the centre of Mercury, others spread out like ripples around the surface. On the opposite side of the planet (at the 'antipode' of the basin), the two sets of shockwaves met, and the resulting tremors pulverized the landscape.

Feature of:	Mercury
Type:	Chaotic terrain shaped by impact shockwaves
Surface area:	250,000 square km (96,000 square miles)
Typical height:	1.8km (1.1 miles)
Age:	4 billion years

Discovery Rupes

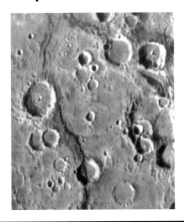

Although Mercury's cratered surface is similar to our Moon, it has some unique features. Most impressive are rupes (from the Latin for 'cliff'), towering vertical cliffs, up to 2km (1.2 miles) high in places, that wind across the landscape separating low-lying areas from raised plateaus. The most impressive is Discovery Rupes, which crosses the middle of a 70km (43-mile) crater, creating a fault down its centre. Astronomers believe that the rupes are linked to Mercury's unusually large core. Excess heat from this molten metal caused the planet to expand early in its history, splitting the crust into separate blocks. As the core cooled, Mercury shrank to smaller than its original size, causing the crust to fall back inwards. Since it was no longer possible for all the pieces to fit, some now stuck out relative to others.

Feature of:	Mercury
Type:	Vertical scarp separating surface regions
Length:	500km (310 miles)
Height:	Up to 2km (1.2 miles)
Age:	About 2 billion years

Venus

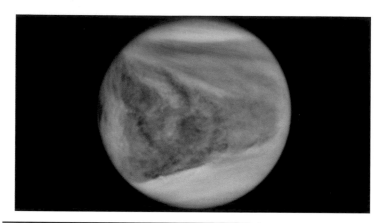

Earth's twin planet in terms of size, and the world closest to our own, Venus is hidden beneath cream-coloured clouds. Astronomers once hoped that these might conceal a lush jungle world, but when the first space probes flew past the planet, they discovered that the planet was scorching hot, and the atmosphere dominated by carbon dioxide. The first robot probes to be landed were destroyed, but revealed clouds of sulphuric acid and a pressure near the surface of 100 Earth atmospheres. In 1975, the Soviet *Venera 9* probe sent back data and images that showed a parched, volcanic world where anything on the surface is simultaneously crushed, boiled and burned. Venus was mapped in detail by the *Magellan* probe, which spent three years orbiting Venus from 1990, equipped with a radar.

Type of object:	Rocky planet
Closest to Sun:	107.5 million km (66.8 million miles)
Furthest from Sun:	108.9 million km (67.6 million miles)
Orbital period:	224 days
Diameter:	12,104km (7518 miles)
Rotation period:	243 days
Axial tilt:	177.4°

Maxwell Montes

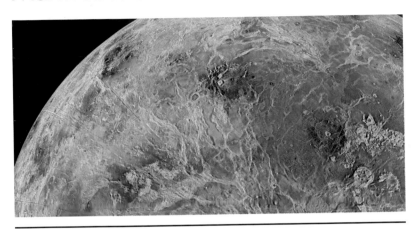

The surface of Venus is dominated by volcanoes, of which the most impressive are the peaks of Maxwell Montes. This highland region, 11km (7 miles) above the average Venusian surface, was discovered by Earth-based radar in the 1960s – it is named after the nineteenth-century physicist James Clerk Maxwell, unlike many features on Venus that are named after historic or mythical females. The mountains sit on a continent-sized plateau called Ishtar Terra, close to the Venusian north pole. The planet's crust never split into tectonic plates like those on Earth, but it has been stretched and compressed by convection in the mantle beneath it. Long trenches mark where it has fallen inwards, while ridges and mountain regions such as Ishtar Terra and Alpha and Beta Regio are formed where it has been pushed upwards.

Feature of:	Venus
Type:	Mountains formed by tectonic compression of surface
Length:	800km (500 miles)
Height:	Up to 11km (7 miles)
Age:	Unknown – probably less than 500 million years

Latona Corona

The volcanoes of Venus are not restricted to the towering cones familiar on Earth. The *Magellan* probe also revealed networks of cracks in the Venusian surface. Some are radiating, spiderlike groups called arachnids. Others are systems of concentric rings called coronae. Both features probably mark places where lava has pushed its way up from below – and once a reservoir of molten rock has exhausted itself, the ground above has collapsed inwards. Volcanic features on Venus are more widespread, and scattered more randomly, than on Earth. This is because Venus has no tectonic plates of the kind that act as a focus for volcanic activity on Earth. With no plate boundaries to allow heat to escape from inside the planet, heat and pressure build up inside until finally 'boiling over' in global eruptions.

Feature of:	Venus
Type:	Volcanic corona bordering Dali Chasma surface fault
Diameter:	1000km (620 miles)
Depth of cracks:	3km (1.9 miles)
Age:	Less than 500 million years

Howe Crater

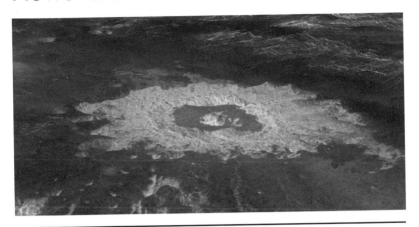

Impact craters are rare on Venus – Howe crater and its siblings Danilova and Aglaonice in the Alpha Regio region are a crater cluster that probably formed from the break-up of a single object in the Venusian atmosphere a few hundred million years ago. The thick atmosphere means that small chunks of interplanetary debris burn up before reaching the ground. The dense air limits how far the spray of ejecta from an impact can spread, and most craters end up with a 'splash' of material around them, rather than the extensive rays seen on airless worlds like the Moon. The pattern of lobes formed by ejecta reveals the angle at which the meteorite approached the ground – the symmetrical pattern around Howe and its siblings shows that they were formed by an impact from almost directly above.

Feature of:	Venus
Type:	Impact crater
Diameter of main crater:	69km (43 miles)
Diameter of ejecta:	160km (100 miles)
Age:	Less than 500 million years

Earth

Our home planet is the largest of the solar system's rocky inner worlds, and third in order from the Sun. With surface water, a hospitable atmosphere and life in abundance, it is unique, thanks to its size and location in the solar system. Earth sits in the middle of the 'habitable zone' around the Sun, where it is possible for liquid water to exist on a planet's surface. A reasonably thick atmosphere, held around the Earth by its fairly strong gravity, prevents the water boiling off into space, and our planet's internal heat has created volcanic activity that splits its crust into drifting plates. The oceans lubricate the plates, and the cycle of water between atmosphere and seas absorbs carbon dioxide and prevents Earth from turning into a Venus-like greenhouse. This makes our planet uniquely hospitable.

Type of object:	Rocky planet
Closest to Sun:	147.1 million km (91.4 million miles)
Furthest from Sun:	152.1 million km (94.5 million miles)
Orbital period:	365.25 days
Diameter:	12,745km (7916 miles)
Rotation period:	23h 56m
Axial tilt:	23.5°

Pacific Ocean

Water covers two-thirds of Earth's surface, and the largest body of water is the Pacific Ocean, occupying almost an entire hemisphere. The oceans fill basins where Earth's crust is at its thinnest – often just a few kilometres deep. The tectonic drift of the crustal plates means that crust is continually being destroyed in some places and formed in others. For example, the edges of the (several) plates that make up the Pacific ocean floor are marked by the 'ring of fire', an area prone to volcanoes and earthquakes where the ocean plate is pushed beneath the continental plates. The plates move a few millimetres in a year, but this changes the pattern of the continents over millions of years. Only the thick central regions of the continents (known as cratons) have survived intact over geological timescales.

Feature of:	Earth
Type:	Ocean basin formed by tectonic drift
Surface area:	169 million square km (65.3 million square miles)
Depth:	Down to 10,911m (35,798ft)
Age:	600 million years

Himalaya Mountains

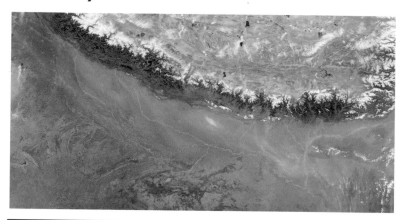

The Himalayas are Earth's most impressive mountain range, and an example of the planet's tectonic plates in action. They are forming as Earth's crust crumples and folds due to the collision between the Indian subcontinent and the bulk of Asia. Until about 10 million years ago, India was an island drifting northwards through the Indian Ocean. Usually a collision between plates forces one or other down into Earth's mantle, but here the ocean floor between the two landmasses crumpled and folded upwards. As the plates continue to converge, the Himalayas rise by about 5mm (0.2in) per year. Not all mountains form in this way – the Earth's other major chain, the Andes, are volcanic, formed by eruptions that have begun as the Nazca Plate of the eastern Pacific plunges beneath the continental plate of South America.

Feature of:	Earth
Type:	Mountain range formed by tectonic collision
Length:	3800km (2400 miles)
Height:	Up to 8.8km (29,000ft)
Age:	About 10–15 million years

Chicxulub Crater

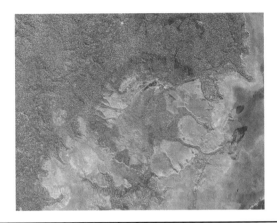

Although Earth's atmosphere protects it from smaller meteorite impacts, it does not offer much shelter from the larger objects that occasionally cross our planet's path. The main reason that craters are not more obvious features of Earth's surface is that weather, water, tectonics and life itself erode and disguise them. Chicxulub, Earth's most infamous crater, is buried beneath the sediments of the Gulf of Mexico near the Yucatan Peninsula. It was discovered in 1978 by geologists searching for oil deposits, and its age – 65 million years old – coincides exactly with the period when the dinosaurs became extinct. The Chicxulub impact scattered debris around the world and may have caused a sudden climate change that helped to kill off the dinosaurs, although biologists are still arguing about its significance.

Feature of:	Earth
Type:	Buried, eroded impact crater
Diameter:	Up to 300km (186 miles)
Depth:	1.3km (4200ft)
Age:	65 million years

Moon

Earth's natural satellite is huge compared to the size of our planet – more than a quarter its diameter. For years astronomers argued over its origins, but rocks brought back by the *Apollo* missions settled the issue – the Moon is a mix of material from Earth and another world that obliterated itself by smashing into our planet 4.5 billion years ago. The Moon is a world of cratered highlands and low-lying, dark seas or maria. Tidal forces have slowed its rotation so that it now spins on its axis in the same time it takes to circle the Earth, meaning that one side always faces our planet, and the other is hidden. The seas are rolling plains of solidified lava, and their lack of craters compared to the highlands suggests they are younger. They formed when lava erupted to flood basins left by impacts 3.9 billion years ago.

Type of object:	Satellite of Earth
Closest to Earth:	363,100km (225,528 miles)
Furthest from Earth:	405,700km (251,988 miles)
Orbital period:	27.32 days
Diameter:	3474km (2158 miles)
Rotation period:	27.32 days
Axial tilt:	1.5°

Sea of Tranquility

The Mare Tranquillitatis is famous as the landing site of the first manned mission to the Moon, *Apollo 11*, in 1969. It was chosen because it could offer an insight into the Moon's geology. The basin in which the 'sea' lies is one of the most irregular and ragged-edged, with several 'bays' along its shores. Even though it was probably formed by a single impact, most signs of its crater walls have been lost because it was one of the first basins to form, and has been overlaid and reshaped by later impacts nearby – such as those in the Mare Nectaris and the Sea of Fertility. Like many lunar seas, the Sea of Tranquility contains 'wrinkle ridges' – low ripples in its surface that indicate places where the basalt lavas that flooded the sea about 3.6 billion years ago formed a skin that crumpled before they froze completely.

Feature of:	The Moon
Type:	Lunar sea
Diameter:	873km (542 miles)
Composition:	Impact basin flooded by basalt lava
Age:	3.6 billion years

Copernicus Crater

A prominent crater on the near side of the Moon, Copernicus is also relatively young, formed 800 million years ago. Terraced cliffs around the edge plunge several kilometres to the crater floor, while three central mountains rise to 1.2km (0.7 miles) in the centre. Copernicus gives its name to an entire geological period of lunar history, sitting at the centre of a system of bright rays and smaller craters formed as its ejecta blasted out across the surface of the Moon to distances of up to 800km (497 miles) from the impact. Because it is quite easy to determine whether a feature formed on top of the ejecta, or was partially covered by it, Copernicus is a keystone to working out the dates of other lunar features; its own age was established from samples of ray material collected by the *Apollo 12* astronauts.

Feature of:	The Moon
Type:	Impact crater
Diameter:	91km (57 miles)
Depth:	3.7 km
Age:	800 million years

Hadley Rille

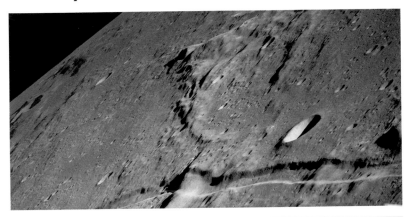

Aside from the solidified lava plains of the maria, the Moon has few signs of volcanic activty. This is probably because its small size meant it cooled rapidly. Until the first probes reached the Moon in the 1950s, astronomers disputed that the Moon's countless craters were volcanic, but the discovery that they range in size down to microscopic made it clear they were caused by impacts. Rilles are the most obvious truly volcanic features – long sinuous valleys flowing out from the volcanic craters and mountains. They are thought to be collapsed lava tubes – underground rivers of lava that remained fluid while the lava plains around them set, and then collapsed as the lava was somehow removed. Hadley Rille is the most impressive, winding through the Lunar Appenine mountains from an outflow near Mons Hadley.

Feature of:	The Moon
Type:	Sinuous rille (collapsed lava tube)
Length:	120km (75 miles)
Depth:	Up to 300m (1000ft)
Age:	about 3.3 billion years

Mars

The Red Planet is the fourth major world from the Sun, much smaller than Earth or Venus, but larger than Mercury. It lies 75 million km (46.6 million miles) further from the Sun than Earth, but it is still the most Earthlike of the other planets. Today it is cold and dry, but the evidence is mounting that it was warmer and wetter in its past, and may have been suitable for life. Mars is tilted on its axis by a similar amount to Earth, and has a rotation period just 30 minutes longer, which gives it Earthlike seasons. However, its orbit is more elliptical, resulting in greater extremes of climate. The Martian surface is split into two halves – sandy lowland plains in the northern hemisphere, and cratered highlands in the south. The distinctive colour comes from large amounts of iron oxides (similar to rust) in the soil.

Type of object:	Rocky planet
Closest to Sun:	206.6 million km (128.3 million miles)
Furthest from Sun:	249.2 million km (154.8 million miles)
Orbital period:	1.88 years
Diameter:	6805km (4226.7 miles)
Rotation period:	24h 38m
Axial tilt:	25.2°

Valles Marineris

This huge scar in the Martian surface is named after the *Mariner 9* probe that found it after entering orbit around Mars in 1971. Grander than Earth's Grand Canyon, it stretches for 4000km (2400 miles), roughly parallel to the equator. Several long channels run in parallel to each other, occasionally opening up into huge basins where the raised ground between them has collapsed entirely. At its deepest, the Mariner valleys plunge to 10km (6 miles) below the surrounding landscape. Unlike the Grand Canyon, it is not caused by water erosion but is a crack in the planet's crust caused by the strain from the Tharsis Bulge to its north (see Olympus Mons, p. 38). Back in the wet Martian past, however, it seems that water did flow into the canyon through the gullies of the Noctis Labyrinthus at its western end.

Feature of:	Mars
Type:	Tectonic canyon system
Length:	4000km (2500 miles)
Width:	Up to 700km (430 miles)
Depth:	Average 8km (5 miles)
Age:	About 3.5 billion years

Olympus Mons

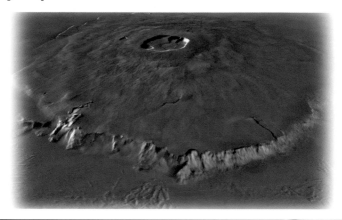

Mars boasts several enormous volcanoes, including this, the largest in the solar system. Towering 27km (17 miles) above the average Martian surface, Olympus Mons is less impressive than its height would suggest, partly because it sits on top of a raised area of crust called the Tharsis Bulge (itself 10km/6 miles high), and partly because it is so wide (500km/311 miles across) that its slopes are very shallow. Nevertheless, its central caldera and its southern edge have cliffs that plunge in drops of more than 1km (0.6 mile). Olympus Mons and its neighbours are shield volcanoes, built up from the eruption of lava over millions of years – in fact, the entire Tharsis Bulge is thought to be a vast volcanic shield. Parts of the volcano's western flank may be as young as two million years old (estimated from the number of craters that have accumulated), suggesting the volcano might still be active.

Feature of:	Mars
Type:	Shield volcano
Diameter:	500km (311 miles)
Height:	27km (17 miles)
Age:	About 100 million years

Polar caps

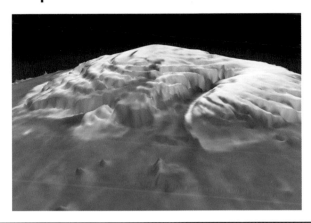

The north and south poles of Mars have their own icecaps – the only places on the planet where water survives on the surface. However, the deposits of ice at the Martian poles are more changeable than those on Earth because they consist of thin layers of water ice, overlaid with larger amounts of frozen carbon dioxide ('dry ice'). As Mars orbits the Sun, and its tilted axis creates a similar cycle of seasons to those on Earth, carbon dioxide frosts settle out of the atmosphere at the beginning of one hemisphere's winter, and evaporate directly back into the air at the onset of spring. The small caps of water ice remain more or less unchanged – there is more water ice on the surface at the north pole than the south, but each pole has large deposits of permanently frozen water mixed with the soil in a permafrost that helps the larger, seasonal carbon dioxide caps form rapidly.

Feature of:	Mars
Type:	Polar ice caps
Diameters:	1100 and 1450km (685 and 900 miles)
Composition:	Seasonal carbon dioxide ice overlaying permanent water ice
Ages:	2.5 billion years or younger

Phobos

Phobos is the larger of two moons in orbit around Mars – its name means 'fear', and it commemorates one of the mythical hounds of Mars, the God of War. The moon is little more than an ovoid lump of rock made of carbon-rich rock, similar to many of the 'C-type' asteroids, and it probably has a similar origin. It seems to be covered in a deep layer of loose soil, or 'regolith', churned up by meteorite impacts throughout its history. Numerous linear scars, whose origins are still not fully explained, run across its surface. The most prominent feature is Stickney, a relatively huge crater about 10km (6 miles) across. Phobos orbits so close to Mars that it is unstable, and will eventually break up to form a ring around the planet, or plunge into the planet's atmosphere to create a new Martian crater.

Type of object:	Satellite of Mars
Closest to Mars:	9236km (5736 miles)
Furthest from Mars:	9519km (5912 miles)
Orbital period:	7h 39m
Diameter:	22.2km (13.8 miles)
Rotation period:	7h 39m
Axial tilt:	Zero

Deimos

The smaller of Mars's moons is similar to Phobos – the chief difference is that Deimos's surface is considerably smoother and less scarred by impacts. Its name means 'panic' – the other of Mars's two hounds. Phobos and Deimos are chemically similar to each other and to the carbon-rich C-type asteroids. They are thought to have begun their lives in the main asteroid belt between Mars and Jupiter, and to have had their orbits disrupted by Jupiter's gravity so they fell into orbits that eventually intersected with Mars and were swept up in orbit around it. The moons were discovered in 1877 by US astronomer Asaph Hall following a deliberate search – their existence had been predicted. Because Deimos orbits quite a bit further from Mars than Phobos, it will not suffer the same destructive fate as its larger sibling.

Type of object:	Satellite of Mars
Closest to Mars:	23,455km (14,569 miles)
Furthest from Mars:	23,465km (14,575 miles)
Orbital period:	1.26 days
Diameter:	12.6km (7.8 miles)
Rotation period:	1.26 days
Axial tilt:	Zero

Ceres

The largest resident of the asteroid belt is now classified as a 'dwarf planet' – a world with enough gravity to pull itself into a spherical shape, but not enough to clear its orbit of other objects. Named in honour of the Greek harvest goddess, Ceres is roughly twice the size of the next largest asteroid, Vesta. No probe has yet visited it (though this should change with the arrival of the *Dawn* mission in 2015), so the best images obtained so far have come from the Hubble Space Telescope. These suggest that Ceres is a heavily cratered world with a dusty outer crust, and has remained largely unchanged since its formation about 4.5 billion years ago. Its most intriguing feature is a bright patch on one side – perhaps marking an area where a meteorite impact has penetrated the dusty crust to reveal a layer of ice.

Type of object:	Asteroid/dwarf planet
Closest to Sun:	381.4 million km (236.9 million miles)
Furthest from Sun:	447.8 million km (278.1 million miles)
Orbital period:	4.6 years
Diameter:	942km (585.1 miles)
Rotation period:	9h 4m
Axial tilt:	Unknown

Vesta

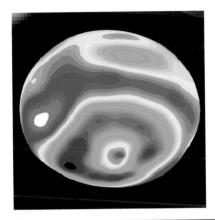

The second largest asteroid is a mystery. Most asteroids have dark surfaces that have changed little since their formation, but Vesta's surface is bright, and covered in material that seems to have erupted in volcanic activity. It is puzzling that this asteroid grew hot enough to power volcanoes when the much larger Ceres did not. The best images of Vesta so far obtained reveal an impact crater gouged out of its southern hemisphere – perhaps the origin of a family of smaller asteroids. The crater also reveals that Vesta's interior has a different composition from its surface, confirming that at one time the asteroid melted all the way through, allowing its minerals to separate according to their density, like those inside a planet. Vesta is a prime target for study by the *Dawn* asteroid probe, which should reach it in 2011.

Type of object:	Asteroid
Closest to Sun:	321.8 million km (199.9 million miles)
Furthest from Sun:	384.7 million km (238.9 million miles)
Orbital period:	3.63 years
Size:	578 x 560 x 458km (359 x 348 x 284 miles)
Rotation period:	5h 21m
Axial tilt:	Unknown

Eros

Eros is one of a group of asteroids that stray sunwards of the main asteroid belt, and which are termed Near Earth Asteroids (NEAs). The second largest NEA, it crosses the orbit of Mars and can approach to within 0.15 Astronomical Units of Earth. Eros was the first asteroid to be intensively studied: the *Near Earth Asteroid Rendezvous* (*NEAR Shoemaker*) probe flew by it in 1998, then entered orbit around Eros, appropriately enough on Valentine's Day, 14 February 2000. The probe studied the cashew-shaped rock for almost a year, gradually orbiting closer to the surface before touching down. One of its discoveries was that many of the asteroid's older craters were erased a billion years ago when shockwaves from an impact shook the surface so much that loose regolith filled in the older features.

Type of object:	Asteroid
Closest to Sun:	169.5 million km (105.3 million miles)
Furthest from Sun:	266.7 million km (165.7 million miles)
Orbital period:	1.76 years
Size:	13 x 13 x 33km (8 x 8 x 20 miles)
Rotation period:	5h 16m
Axial tilt:	Unknown

Gaspra

The first asteroid to be visited by a probe, Gaspra orbits close to the inner edge of the main asteroid belt, and was photographed from a distance of about 5000km (3100 miles) by the Jupiter-bound *Galileo* probe in 1991. Images sent back to Earth revealed an irregularly shaped rock with small impact craters across its surface. Although there are various dents in the asteroid's shape, none has been identified beyond doubt as an impact crater, and this suggests that Gaspra's current surface has been exposed to space for only a relatively brief time. While most asteroids are carbon-rich 'C types', Gaspra is a silicate rich 'S type'. Its surface minerals link it to a group of asteroids known as the Flora family. If this family formed from the breakup of a single larger world, this might explain why Gaspra's surface is relatively fresh.

Type of object:	Asteroid
Closest to Sun:	273 million km (169.6 million miles)
Furthest from Sun:	388.1 million km (241.1 million miles)
Orbital period:	3.28 years
Size:	18 x 11 x 9km (11 x 7 x 6 miles)
Rotation period:	7h 3m
Axial tilt:	Unknown

Mathilde

The *NEAR Shoemaker* probe flew by Mathilde in 1997. Photos of this misshapen rock, roughly 66km (41 miles) across, show a world that has been hammered into its current shape by impacts from all sides, yet whose craters are surprisingly sharp-edged, lacking the usual blanket of ejecta caused by material flung out by the impact settling back on to the surrounding landscape. Data from the flyby confirmed that Mathilde has extremely weak gravity – too feeble to hold on to the rubble from impacts on its surface – and an average density close to that of water. Since the visible surface seems solid enough, the only explanation must be that the asteroid's interior contains huge voids of empty space. It now seems that many asteroids, like Mathilde, are little more than orbiting heaps of loosely attracted rubble.

Type of object:	Asteroid
Closest to Sun:	290.6 million km (180.5 million miles)
Furthest from Sun:	501.3 million km (311.4 million miles)
Orbital period:	4.31 years
Size:	66 x 48 x 46km (41 x 30 x 29 miles)
Rotation period:	17.41 days
Axial tilt:	Unknown

Ida and Dactyl

The first asteroid to be discovered with its own satellite, Ida was only the second satellite to be studied up close, when the *Galileo* probe flew past it in 1993. Like Gaspra, Ida is an 'S-type', silicate-rich asteroid, but analysis of its spectrum from Earth puts it in a group called the Koronis family. These are fragments of a larger asteroid believed to have broken up relatively recently, but the discovery that Ida's surface was relatively heavily cratered has pushed back the date of this break-up. Dactyl was discovered only when Galileo's images were properly analyzed in 1994. It is a tiny world just 1.4km (0.9 mile) across, held in orbit by Ida's weak gravity. Astronomers are unsure whether it is a captured body, a world that formed alongside Ida, or a fragment of the larger asteroid which was knocked into orbit.

Type of object:	Asteroid (and moon)
Closest to Sun:	408.2 million km (253.5 million miles)
Furthest from Sun:	447.8 million km (278.1 million miles)
Orbital period:	4.84 years
Size:	54 x 24 x 15km (34 x 15 x 9 miles)
Rotation period:	4h 37m
Axial tilt:	Unknown

Jupiter

Jupiter is the largest planet – large enough to contain all the other planets with room to spare. It is a gas world, with layers of gas (dominated by hydrogen and helium) that may or may not surround a small solid core. At more than 7000km (4350 miles) below the surface, the gas is compressed into liquid form, and deeper inside, hydrogen molecules split into atoms to form an ocean of liquid metallic hydrogen, which generates a magnetic field. Jupiter spins rapidly, in just 10 hours, and its equatorial regions move so fast that they bulge outwards. The upper layers of the atmosphere consist of complex weather systems wrapped around the planet by its rapid spin to create bands of high and low pressure, marked by clouds of varying colours. Light areas are referred to as zones and darker areas as belts.

Type of object:	Gas giant planet
Closest to Sun:	740.6 million km (460 million miles)
Furthest from Sun:	816.5 million km (507.1 million miles)
Orbital period:	11.9 years
Diameter:	142,984km (88,810 miles)
Rotation period:	9h 56m
Axial tilt:	3.1°

Great Red Spot

Jupiter's most famous feature is the Great Red Spot (GRS), an anticyclone three times the size of Earth, in the planet's southern hemisphere. Here, very low air pressure draws chemicals from deep within the atmosphere to about 8km (5 miles) above the surrounding cloud layers, where they condense to form reddish clouds. The spot has lasted for centuries – astronomers have tracked its movements since at least 1830, and there are inconsistent reports dating back to 1655. It is the grandest of Jupiter's many storms – less intense ones form white spots that drift between the cloud bands, growing, fading and occasionally merging together. In the late 1990s, three spots formed 50 years previously merged to form a single superstorm that has since intensified and changed colour to form a rival 'Little Red Spot'.

Feature of:	Jupiter
Type:	Anticyclone weather system
Dimensions:	24–40,000 x 12–14,000km (15–25,000 x 7–9000 miles)
Height:	8km (5 miles) above surrounding atmosphere
Age:	At least 200 years (perhaps 400 or more)

Jupiter's rings

Like all the giant planets, Jupiter has its own ring system – though nothing to rival those around Saturn or Uranus. It was discovered only in 1977, when the *Voyager 1* probe turned its cameras to photograph the retreating dark side of Jupiter following its closest flyby, and captured a thin plane of dust backlit by the Sun – from the Earth's side of Jupiter, the rings had been drowned out by the glare of Jupiter's sunlit face. We now know that Jupiter has four rings of varying density and composition, known (from Jupiter outwards) as the Halo, the Main Ring, and the Amalthea and Thebe Gossamer rings. Each is formed from particles that are continuously blasted off Jupiter's inner moons – Metis and Adrastea, Amalthea and Thebe – by the impacts of micrometeorites pulled in by Jupiter's immense gravity.

Ring components:	Halo ring, main ring, Amalthea Gossamer Ring, Thebe Gossamer Ring
Overall radius:	92,000–226,000km (57,000–140,000 miles)
Typical thickness	About 30km (19 miles)
Size of particles	0.001mm
Composition:	Dust particles

Amalthea

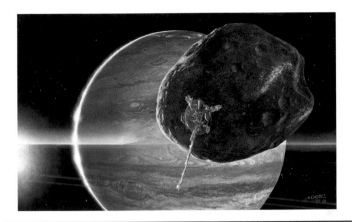

This small, elongated world, the outermost of Jupiter's inner moons, looks like a captured asteroid, but the composition of its rocks suggests that it is a fragment of a larger moon that broke up – after an interplanetary collision or through the stresses exerted by Jupiter's powerful gravity. What survives is one of the reddest worlds in the solar system, with a green tinge on parts of the surface. Amalthea is constantly bombarded by micrometeorites that blast dust particles from its surface. As these spiral towards Jupiter, they form the Amalthea Gossamer Ring. One explanation for the green material on the moon's surface is that the moon sweeps up sulphur-rich material escaping from Io, and the effects of radiation this close to Jupiter alter the sulphur compounds, changing their colour from yellow to green.

Type of object:	Satellite of Jupiter
Closest to Jupiter:	181,150km (112,516 miles)
Furthest from Jupiter:	182,840km (113,565 miles)
Orbital period:	11h 57m
Diameter:	250 x 146 x 128km (155 x 91 x 80 miles)
Rotation period:	11h 57m
Axial tilt:	Zero

Io

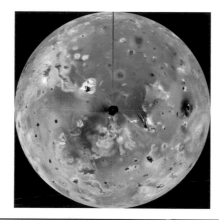

The innermost of Jupiter's four 'Galilean' moons, Io is also the second smallest – though still larger than Earth's Moon. Orbiting close to Jupiter, it is tortured by the planet's gravity, which produces enormous tides that pummel the moon's interior and keep it molten. Heat escaping through the surface makes Io the most volcanic world in the solar system. Pools of molten sulphur seethe in volcanic calderas on the surface, while liquid sulphur-based compounds erupt in geysers that evaporate into plumes of vapour mushrooming above the surface. Sulphur and its compounds give Io its lurid appearance – it has been described as resembling a burnt, mouldy pizza. This volcanic activity also ensures that Io is the world with the fastest-changing surface in the solar system, reshaping itself completely in decades.

Type of object:	Satellite of Jupiter
Closest to Jupiter:	420,000km (260870 miles)
Furthest from Jupiter:	423,400km (262981 miles)
Orbital period:	1.77 days
Diameter:	3643km (2263 miles)
Rotation period:	1.77 days
Axial tilt:	Zero

Pele

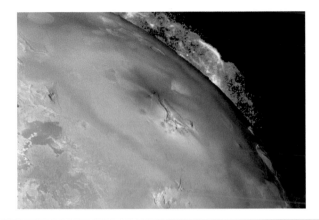

The discovery of volcanic activity on Io was a surprise to the scientists monitoring the first close flyby of the moon, by *Voyager 1* in 1977. It was only as the cameras turned back to look at Io's retreating night side that they saw a mushroom cloud of condensing sulphur droplets hanging 300km (186 miles) above the moon's limb. More volcanic plumes were soon discovered, and later probes traced the first to a volcanic caldera on the surface named Pele (after the Hawaiian volcano goddess). Pele was active during *Galileo*'s mission to Jupiter in the 1990s, and is surrounded by a red ring of sulphur compounds that have fallen back on to the surface. Io's eruptions are often so powerful that they fling clouds of sulphur out of Io's gravity altogether, creating a doughnut-shaped cloud of vapour around Io's orbit.

Feature of:	Io (satellite of Jupiter)
Type:	Volcano with associated sulphur plume
Diameter:	30 x 20km (18 x 12 miles)
Height of plume:	Up to 300km (180 miles)
Age:	Unknown

Europa

The smallest of the Galilean moons, and second in order from Jupiter, Europa is a very different world from Io. A ball of brilliant ice, it is mottled with pink and brown stains and scarred with tracklike markings. Despite such markings, Europa is the smoothest world in the solar system – scaled up to the size of Earth, it would have no hills higher than a few hundred metres. Visiting probes have established that the surface is just a crust of ice several kilometres thick, covering a global ocean of liquid water. The interior of Europa is kept from freezing solid by heat from the moon's interior, which takes a similar tidal pounding to Io. Europa's ocean floor probably resembles some of Earth's volcanic deep-ocean ridges. Io's icy surface is just fluid enough to flow and flatten out any peaks and troughs that form.

Type of object:	Satellite of Jupiter
Closest to Jupiter:	664,300km (412,609 miles)
Furthest from Jupiter:	677,900km (421,056 miles)
Orbital period:	3.55 days
Diameter:	3121.6km (1938.9 miles)
Rotation period:	3.55 days
Axial tilt:	Zero

Tyre

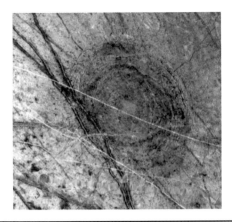

Impact craters are rare on Europa – the mobility of the surface destroys them quite quickly. The largest surviving impact feature so far identified is the multi-ringed Tyre. The impact that formed it was relatively small, and the original crater sat neatly inside the innermost ring visible today. The concentric rings seem to have formed because the crater went deep enough to create a hole in the crust. As water from below rushed in to fill the gap, it dragged the crust above it inwards, causing it to fracture in concentric circles. Analyzing the size and shape of craters has let scientists estimate that the icy crust above Europa's liquid ocean is about 19–25km (12–16 miles) thick. Cracks and fissures in the surface seem common enough to allow water to well up and create the various tracks and stains on Europa's surface.

Feature of:	Europa (satellite of Jupiter)
Type:	Impact crater with surrounding 'ripple' rings
Diameter:	40km (25 miles) for central crater
Relief:	A few tens of metres
Age:	Unknown

Ganymede

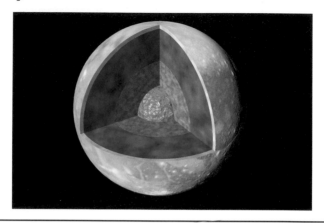

The largest moon in the solar system, Ganymede is the third of Jupiter's 'Galilean' moons, far enough away from its parent planet that it does not suffer from large tidal forces. Probes have revealed Ganymede as a patchwork world of light and dark terrains. It seems that at some point in its past, Ganymede had a different orbit, in which it suffered considerable tidal heating, which warmed up its interior of icy rock and let the moon develop its own version of the plate tectonics seen on Earth. Sections of older crust split and drifted apart, occasionally sinking below the surface, while a slushy mixture of rock and ice welled up to fill in the gaps. Although the surface has long since refrozen, remaining internal heat could explain one of its most intriguing features – a large ocean of water beneath the solid crust.

Type of object:	Satellite of Jupiter
Closest to Jupiter:	1,069,200km (664,099 miles)
Furthest from Jupiter:	1,071,600km (665,590 miles)
Orbital period:	7.15 days
Diameter:	5262.4km (3268.6 miles)
Rotation period:	7.15 days
Axial tilt:	Zero

Uruk Sulcus

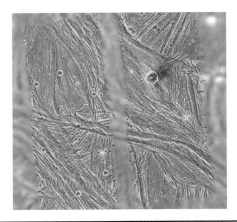

Orbiting among Jupiter's moons for a few years, the *Galileo* probe photographed the traces of Ganymede's tectonic activity and found several features called sulci – regions of parallel grooves in the surface, stretching for tens, even hundreds of kilometres. One of the most spectacular is Uruk Sulcus, where several grooved regions criss-cross each other. Scientists calculated the age of these 'units' by counting how many craters they have accumulated, which reveals how long they have been exposed to space. Most sulci are a mix of older and younger surfaces, and three-dimensional images created by studying the area from different angles reveal that the sulci are areas where parallel faults opened in Ganymede's surface, and older areas slumped or toppled over, with ice from within welling up to fill the caps.

Feature of:	Ganymede (satellite of Jupiter)
Type:	Ridged terrain formed by tectonic extension
Image area:	Approx. 60km (37 miles) across
Relief:	250–500m
Age:	Older than 1 billion years

Callisto

Callisto, the outermost and second largest of Jupiter's Galilean moons, differs from its siblings. All have been shaped by geological activity since their formation, but Callisto has changed little, protected from the effects of tidal heating by its distance from Jupiter. A deep-frozen ball of rock and ice, it is shaped only by the influence of meteorite impacts, and may be the most cratered body in the solar system (since Jupiter's gravity pulls asteroids and comets into its path). The result is a planetary glitterball in which dark, ancient terrain is pockmarked with impact craters that expose brighter ice from below the surface, and spray it across the landscape in bright rays. In places, debris embedded in the surface has been eroded by millions of years of weak sunlight, forming landscapes of eerie conical spires.

Type of object:	Satellite of Jupiter
Closest to Jupiter:	1,869,000km (1,160,870 miles)
Furthest from Jupiter:	1,897,000km (1,178,261 miles)
Orbital period:	16.69 days
Diameter:	4820.6km (2994.2 miles)
Rotation period:	16.69 days
Axial tilt:	Zero

Valhalla

Callisto's most impressive feature is the Valhalla impact basin, the remains of a meteorite or comet impact up to four billion years ago. The event that formed Valhalla was one of a handful of impacts through Callisto's history large enough to have punctured the outer crust completely, letting icy material well up from inside the moon to form a bright area known as a palimpsest. Astronomers assumed for years that Callisto had an essentially uniform interior, and that the crust was darker only because of its exposure to space – but there is evidence that, like Europa and Ganymede, it may conceal a layer of inexplicably liquid water below its surface. The *Galileo* probe discovered this ocean from the weak magnetism it generates as the moon ploughs through Jupiter's own, far more powerful, magnetic field.

Feature of:	Callisto (satellite of Jupiter)
Type:	Impact crater and surrounding ridges
Diameter of central basin:	600km (370 miles)
Diameter including outer ridges:	About 2600km (1600 miles)
Scale of ridges:	About 50km (30 miles) apart
Age:	About 3 billion years or older

Saturn

The second largest world in the solar system, Saturn is famed for its ring system*, while the planet itself is often overlooked. Saturn is a gas giant similar to Jupiter and with a similar composition, but it has none of its sibling's colourful cloud patterns – instead it is sepia toned. This is due to several factors. Most significant is the difference in its mass – it weighs less than a third as much as Jupiter, but its weaker gravity allows its outer layers to expand and gives it the lowest density of any planet – less than that of water. Lower gravity means that the equatorial bulge caused by its rapid rotation is even more pronounced than on Jupiter. Meanwhile, the lower temperatures further from the Sun allow ammonia from the atmosphere to condense and form a haze that obscures features in the clouds below.

Type of object:	Gas giant planet
Closest to Sun:	1353 million km (840.4 million miles)
Furthest from Sun:	1513 million km (939.8 million miles)
Orbital period:	29.7 years
Diameter:	120,536km (74,867 miles)
Rotation period:	10h 32m
Axial tilt:	26.7°

White Spots

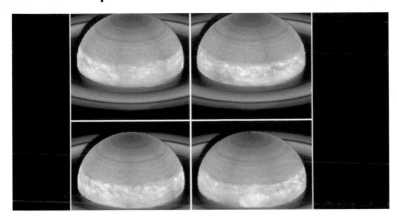

The largest storms on Saturn are formed by white ammonia clouds condensing at altitude in areas of low pressure. False-colour images reveal various colours among the weather features, indicating different chemicals condensing at various temperatures and pressures in the atmosphere. Storms are short-lived; Saturn's low density may help to disperse them. Great White Spots form once in every Saturnian year (30 Earth years), as its northern hemisphere goes through midsummer. Forces generated by Saturn's rapid rotation stretch the storms out, transforming them into bands around the equator, which finally disperse. The weather systems are powered by energy from the Sun and by heat from inside the planet, where it is released by chemical changes and the sifting of denser and lighter materials.

Features of:	Saturn
Type:	Low-pressure atmospheric storms
Typical diameters:	Up to 21,000km (13,000 miles)
Typical durations:	Several months
Prominent appearances:	1876, 1903, 1933, 1960, 1990, 1994, 2006

Saturn's rings

The rings of Saturn are not solid. Rather, they are planes made up of countless ringlets, each a circle of individual particles of varying sizes, kept in order by the need to avoid colliding with their neighbours. Different sizes of particle, and the gravitational influence of various moons, create broad rings with comparatively empty divisions between them. The brightest rings seen from Earth are the outer A and inner B rings, separated by the Cassini Division. Saturnward of the B ring lie the more transparent C and D rings. Beyond the A ring (itself split in two by the narrow Encke Division) lies the thin braid of the F ring and the broad but sparse E ring. The rings lie above Saturn's equator, and we see them from different angles as Saturn moves through its 29-year orbit.

Main ring components:	D Ring, C Ring, B Ring, Cassini Division, A Ring, Encke Division, F Ring, G Ring, E Ring
Overall radius:	66,900–483,000km (41,500–300,000 miles)
Typical thickness	About 100m
Size of particles	A few centimetres > 5m (a few inches > 16ft)
Composition:	Water ice?

Moonlets

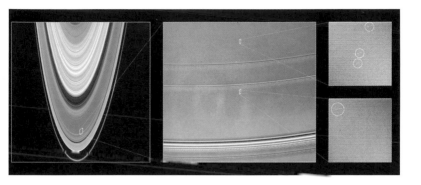

Saturn's rings may have been formed by the break-up of a comet or moon close to Saturn. As the debris particles spread out in orbit around the planet, countless collisions flattened their orbits and forced them into circles. However, the rings are unstable and particles from the semi-transparent C and D rings are constantly sifting down into Saturn's atmosphere. To keep the rings replenished, larger fragments of rock must be continuously breaking down, probably as 'moonlets' in the outer rings are bombarded by micrometeorites. The gravity of these moonlets keeps the rings in order, and creates propeller-shaped gaps in the ring structure, as well as short-lived radial 'spokes'. The ring structure arises because particles straying into other orbits will be ejected from the system or shepherded into line.

Feature of:	Saturn's rings
Type:	Small moonlets generating fine ring structure
Moonlet diameter:	100m (300ft)
'Propeller length':	500m (1,500ft)
Age:	Tens of millions of years

Pandora

Pandora is one of several 'shepherd' moons that orbit around Saturn's F Ring: Atlas and Prometheus on the ring's inside edge, Pan slightly closer to Saturn and Pandora on the outer edge. Any particles straying out of their circular orbits around Saturn and into a more elliptical path would probably encounter one of the moons, and be yanked back into line by its gravity, smashed into the surface of the moon, or ejected from the Saturnian system altogether. Each moon is 100–200km (62–124 miles) in diameter, and only roughly spherical. They can affect particles in more distant orbits that share 'resonances' with them – for example, where a particle orbits in precisely two-thirds of the period of a moon, it will suffer repeated encounters with the moon's gravity that can create a clearing among the ringlets.

Type of object:	Satellite of Saturn
Closest to Saturn:	141,125km (87,655 miles)
Furthest from Saturn:	142,315km (88394 miles)
Orbital period:	15h 4m
Size:	103 x 80 x 64km (64 x 50 x 40 miles)
Rotation period:	15h 4m
Axial tilt:	Zero

Epimetheus

A rare example of a 'co-orbital' moon, Epimetheus shares its orbit around Saturn with another moon, Janus. The two act as 'shepherds' to the F ring, fulfilling a similar role to Pandora and its siblings. They are thought to be fragments of a larger object that broke apart in orbit around Saturn. Most of the time, one moon orbits closer to Saturn than the other, but every four years they catch up with each other, exchanging momentum due to their gravitational pull on each other. This increases the speed of the outer moon and moves it closer to Saturn, while slowing down the inner moon and causing it to drift further out. Eventually, the moons swap positions. Astronomers spotted Epimetheus and Janus in the 1960s, when the rings lay edge-on to Earth, and their existence was confirmed by the *Voyager* flybys of the 1970s.

Type of object:	Satellite of Saturn
Closest to Saturn:	149,926km (93,122 miles)
Furthest from Saturn:	152,894km (94,965 miles)
Orbital period:	16h 40m
Size:	135 x 108 x 105km (84 x 67 x 65 miles)
Rotation period:	16h 40m
Axial tilt:	Zero

Mimas

The innermost of Saturn's major moons, Mimas is a small world, just 418km (260 miles) across. This far out in the solar system, satellites tend to be a mix of rock and ice, and Mimas is no exception. Heavily cratered, its surface seems to have seen little activity since it formed more than four billion years ago. One hemisphere is dominated by a crater called Herschel, named after William Herschel, who discovered Mimas in 1789. Herschel may be the biggest crater in proportion to the size of its host world in the entire solar system – 130km (81 miles) across and 10km (6 miles) deep, it must have been formed by an impact that shook Mimas to the core. Even though it orbits beyond Saturn's rings, Mimas seems not to have been influenced by tidal heating – unlike its outer neighbour Enceladus.

Type of object:	Satellite of Saturn
Closest to Saturn:	181,659km (112,832 miles)
Furthest from Saturn:	189,149km (117,484 miles)
Orbital period:	22h 37m
Size:	414 x 394 x 381km (257 x 245 x 235 miles)
Rotation period:	22h 37m
Axial tilt:	Zero

Enceladus

Orbiting beyond Mimas, Enceladus is a surprise – a small moon that should be frozen solid like its inner sibling, but is instead an active world of snow, geysers and perhaps even subsurface rivers. An early hint that Enceladus was unusual came when the *Voyager* probes discovered it has the most reflective surface of any world in the solar system. Astronomers were soon speculating that its pristine and relatively craterless surface is continually refreshed by snowfalls from geysers beneath the surface, but not until the *Cassini* probe in 2005 did they realize the extent of activity: a flyby detected a plume of icy particles rising high above it, later photographing it in silhouette. Some particles escape from Enceladus altogether and spread out inwards of its orbit to form Saturn's tenuous E Ring.

Type of object:	Satellite of Saturn
Closest to Saturn:	236,830km (147,099 miles)
Furthest from Saturn:	239,066km (148,488 miles)
Orbital period:	1.37 days
Diameter:	504.2km (313.2 miles)
Rotation period:	1.37 days
Axial tilt:	Zero

Tiger Stripes

Enhanced colour images of Enceladus reveal bluish stripes on the surface of its southern hemisphere. These 'tiger stripes' indicate areas where the surface is warmer than elsewhere – around -180°C (-290°F), compared to an average of -200°C (-326°F) – and seem to correspond to areas that receive the maximum tidal heating and stretching thanks to the gravitational tug from the larger outer moons. It is also the origin of most of Enceladus's ice geysers, which suggests that streams of water run just below the surface here, breaking through and boiling off into space at weak points. In contrast to the global oceans suspected on several of Jupiter's moons, Enceladus's water seems to be more localized – it is probably melted by volcanic activity. The entire moon may also benefit from a rockier composition than its neighbours, meaning it contains more heat-producing radioactive elements.

Feature of:	Enceladus
Type:	Cryovolcanic regions of weakened ice crust
Typical length:	130km (80 miles)
Typical separation:	40km (25 miles)
Age:	10–1000 years

Tethys

The next two major moons in order from Saturn are a step up in size – Tethys and Dione are both roughly 1000km (620 miles) across. The inner of the two, Tethys, is comparatively bright and icy, though more heavily cratered than Enceladus. A crater called Odysseus scars one side of the planet, but has flattened out considerably since it formed, probably due to the slow flow of the icy crust. A fault called Ithaca Chasma runs for hundreds of kilometres around the planet, far away from Odysseus but parallel to its edge, and was probably opened up by the gradual slipping of the terrain. Although Tethys has many craters, parts of it seem to have been resurfaced in its distant past, and there is other evidence of rock and ice flowing on Saturn's moons in the same way that lava does on warmer worlds.

Type of object:	Satellite of Saturn
Closest to Saturn:	294,619km (182,993 miles)
Furthest from Saturn:	294,619km (182,993 miles)
Orbital period:	1.88 days
Diameter:	1066km (662.1 miles)
Rotation period:	1.88 days
Axial tilt:	Zero

69

Dione

Dione is a world with two distinct sides – one that is comparatively bright and less cratered, the other that is darker, with more craters and bright wisps criss-crossing it. As on Tethys, many of Dione's craters have slumped and flattened out since they formed, implying that the icy surface is able to flow very slowly. In 2005, the *Cassini* probe discovered that the 'wispy terrain' consisted of networks of bright-faced cliffs – now frozen solid, Dione's surface was warm enough in its early history to drift and split apart . Elsewhere, icy 'lava flows' have wiped away the traces of ancient craters – astronomers believe that this icy 'cryovolcanism' happens where water ice and ammonia mix together to form a slush with an extremely low melting point, which can be warmed by even weak tidal forces.

Type of object:	Satellite of Saturn
Closest to Saturn:	376,566km (233,892 miles)
Furthest from Saturn:	378,226km (234,923 miles)
Orbital period:	2.74 days
Diameter:	1123.4km (697.8 miles)
Rotation period:	2.74 days
Axial tilt:	Zero

Rhea

The next major moon of Saturn is another step up in size from Tethys and Dione, though its composition is similar. However, Rhea shows few signs of the activity seen on its inner neighbours but is covered in a dense blanket of craters (and a few ice cliffs) that seem to have changed little since the moon first formed. Rhea's additional size means it must have formed with extra internal heat, but scientists think that the moon's greater density worked against it, compressing its water-ice interior so much that it could no longer interact with ammonia to form liquid slush. As a result, cryovolcanism on Rhea never had a chance to begin. Despite this, many of the larger craters have bright icy rims that were probably formed by bright subsurface ice that escaped to the surface after being melted by the heat of impact.

Type of object:	Satellite of Saturn
Closest to Saturn:	526,444km (326,984 miles)
Furthest from Saturn:	527,772km (327,809 miles)
Orbital period:	4.52 days
Diameter:	1528.6km (949.4 miles)
Rotation period:	4.52 days
Axial tilt:	Zero

Titan atmosphere

Saturn's largest moon by far, Titan was discovered by Dutch astronomer Christiaan Huygens in 1655. It was long thought to be the largest satellite in the entire solar system, and a flyby was seen as a priority for the *Voyager* interplanetary probes launched in the mid-1970s. But *Voyager* images of Titan proved frustrating – the moon's atmosphere, whose existence was already suspected, proved to be so dense and opaque that the surface was hidden from view. Titan is unique among the solar system's moons in possessing a substantial atmosphere, which it can hold on to because of its comparatively strong gravity, and because gases move more sluggishly this far from the Sun. One of the few definite discoveries was that Titan was slightly smaller than Ganymede – its atmosphere just made it look larger.

Type of object:	Satellite of Saturn
Closest to Saturn:	1,186,680km (737,068 miles)
Furthest from Saturn:	1,257,060km (780,783 miles)
Orbital period:	15.95 days
Diameter:	5152km (3200 miles)
Rotation period:	15.95 days
Axial tilt:	Zero

Titan – beneath the clouds

In 2005, the *Cassini* spacecraft arrived in orbit around Saturn, carrying Huygens, a probe designed to land on Titan, and infrared imaging cameras capable of looking through Titan's atmosphere and photographing the surface below. The subsequent images revealed an Earthlike world, largely shaped by erosion. This confirms that conditions on Titan let methane exist in solid, liquid and vapour forms, cycling through the atmosphere just as water cycles between water vapour, rain, rivers, seas, and ice caps on Earth, eroding the landscape. It seems, too, that Titan is warm enough for icy volcanism of the type that once shaped Tethys and Dione to continue, wiping the surface clean and leaving just a few craters on its surface. Sites of recent volcanic activity are marked by Enceladus-like 'hot spots' on the moon's surface.

Feature of:	Titan (satellite of Saturn)
Type:	Surface lakes
Image type:	False-colour radar map (smooth areas shown dark)
Typical diameters:	A few kilometers to tens of kilometers
Diameter:	9758.8km (6061.4 miles)
Composition:	Liquid methane

Titan's surface

The *Huygens* lander plunged into Titan's atmosphere on 14 January 2005, carrying instruments that gave us our first direct look at the surface. As *Huygens* emerged from the haze into the much clearer lower atmosphere, it deployed its main parachutes and began to drift down into what looked like a bay with a coastline eroded by rivers and an offshore island. Huygens eventually touched down in a river 'estuary' among rock/ice pebbles that seemed to have been dropped by the river as it emptied out. The surface was covered in a brittle layer that mission scientists interpreted as frost. One disappointment of the mission was that *Huygens* did not directly detect methane as either rain or surface liquid. However, it seemed that the lander had just missed a 'shower', and later observations by the orbiting *Cassini* probe have confirmed lakes of liquid methane elsewhere on Titan.

Feature of:	Titan
Type:	River outflow?
Height of camera:	40cm (17in) above surface
Surface composition:	Frozen water ice with impurities and methane permafrost
Surface temperature:	-179°C (-290°F)

Hyperion

Orbiting beyond Titan, and thus not influenced by its gravity, Hyperion is one of the strangest worlds in the solar system – an oddly shaped satellite with a spongelike surface and an unpredictable rotation period. Since Hyperion should be large enough to have pulled itself into a spherical shape, it may be all that remains of a larger world that suffered a catastrophic collision in its past. Astronomers do not know how its honeycomb surface was created, but it is probably the result of aeons of sunlight evaporating some more easily melted areas, while seams of more durable material have remained unchanged. Analyzing the colours of Hyperion's surface has revealed chemical signatures of water and carbon dioxide ice, and suggests that many craters are filled with dark carbon-based chemicals.

Type of object:	Satellite of Saturn
Closest to Saturn:	1,298,836km (806,730 miles)
Furthest from Saturn:	1,663,182km (1,033,032 miles)
Orbital period:	21.28 days
Size:	360 x 280 x 225km (224 x 174 x 140 miles)
Rotation period:	Chaotic
Axial tilt:	Variable

Iapetus

Another bizarre Saturnian moon, Iapetus is strange even when observed from Earth, since it changes its apparent brightness from one side of its orbit to the other. After discovering it in 1671, Italian astronomer Giovanni Cassini realized that Iapetus must be tidally locked with the same face pointing towards Saturn, and that its 'leading' hemisphere (the one that faces forward as it moves along its orbit) was much darker than the opposite, 'trailing' hemisphere. This was confirmed by the *Voyager* and *Cassini* probes. Photographs of craters also revealed that Iapetus is essentially a light moon with a dark coating, rather than the other way around. Most astronomers believe that Iapetus has picked up its coating of dark material as it flies into debris strewn along its orbit, but the origin of this debris is still uncertain.

Type of object:	Satellite of Saturn
Closest to Saturn:	3,458,936km (2,148,408 miles)
Furthest from Saturn:	3,662,704km (2,274,971 miles)
Orbital period:	79.32 days
Diameter:	1471km (914 miles)
Rotation period:	79.32 days
Axial tilt:	Zero

Iapetus equatorial ridge

One of the strangest features on Iapetus, discovered by the *Cassini* spaceprobe, is a straight ridge that runs around the equator, looking remarkably artificial. The ridge is on average 20km (12 miles) wide and 13km (8 miles) high, and stretches for at least 1300km (808 miles) around Iapetus. Astronomers are not sure how the ridge formed – it may consist of material that was somehow dumped on to the moon's surface, or it may be made from icy material that welled up from inside the moon and then 'migrated' towards the equator under the influence of forces from Iapetus' rotation. A third idea is that Iapetus once span much faster – fast enough to develop a pronounced equatorial bulge. As the moon cooled and its rotation was slowed by tides from Saturn, the ridge section of the bulge somehow remained in place even as the rest of the moon returned to a more spherical shape.

Feature of:	Iapetus (satellite of Saturn)
Type:	Raised region of surface
Length:	At least 1300km (800 miles)
Width:	20km (12.5 miles)
Height:	13km (8 miles)
Age:	Unknown (probably greater than 3 billion years)

Phoebe

Beyond Iapetus, Saturn is surrounded by a swarm of small outer moons – captured comets or asteroids in long elliptical orbits. Phoebe sits awkwardly between the large inner moons and the small outer ones. With an average diameter of 220km (137 miles), it is larger than any of Saturn's other outer moons, yet it goes the 'wrong way' around Saturn compared to all the inner moons (which would have formed from the same cloud of gas and dust as the planet itself). In 2004, *Cassini* solved a few of Phoebe's mysteries as it flew past on its way into the Saturnian system – it found that the moon was an irregularly shaped ball with a dark surface, deep craters and a more rocky composition than Saturn's natural satellites. This suggests that Phoebe is a large captured comet or 'centaur' such as Chiron.

Type of object:	Satellite of Saturn
Closest to Saturn:	10,931,532km (6,789,771 miles)
Furthest from Saturn:	14,979,986km (9,304,339 miles)
Orbital period:	550.56 days
Diameter:	230 x 220 x 210km (143 x 137 x 130 miles)
Rotation period	9h 16m
Axial tilt:	152.1°

Chiron

Discovered in 1977 but since traced in images dating back to 1895, Chiron was originally classed as a rogue asteroid, on an elliptical path from just inside Saturn's orbit at its closest to the Sun, to just inside Uranus's at its most distant. As similar objects were found, astronomers realized there was a family of such worlds in the outer solar system. They called them centaurs – distant equivalents of the Near Earth Asteroids whose orbits overlap the inner planets (Chiron was a wise centaur in Greek mythology). Their origin is puzzling – their surfaces have differing colours and a handful, including Chiron, develop comet-like atmospheres close to the Sun. They may be objects from the Kuiper Belt that have fallen towards the Sun, but will eventually become short-period comets in closer orbits.

Type of object:	Centaur (satellite of Saturn)
Closest to Sun:	1263 million km (784.5 million miles)
Furthest from Sun:	2826 million km (1755.3 million miles)
Orbital period:	50.5 years
Diameter:	137km (85.1 miles)
Rotation period:	5h 55m
Axial tilt:	Unknown

Uranus

The first planet discovered with a telescope, Uranus was found by William Herschel in 1781, though it is visible to the naked eye if one knows where to look. The planet is much smaller than Jupiter or Saturn, and is sometimes classified, along with Neptune, as an ice giant. Beneath its layers of gas lies a deep mantle of chemical ices, including water, methane and ammonia, surrounding an Earth-sized rock/ice core. The upper gas layers are composed of hydrogen and helium, but get their distinctive greenish colour from a small amount of methane, which absorbs red light. Uranus has 27 known moons and a system of rings second only to Saturn's, but its strangest feature is its extreme tilt – at 98° from vertical, the planet literally rolls around its orbit, resulting in a bizarre cycle of days and seasons.

Type of object:	Ice giant planet
Closest to Sun:	2749 million km (1707.5 million miles)
Furthest from Sun:	3004 million km (1865.8 million miles)
Orbital period:	84.3 years
Diameter:	51,118km (31,750 miles)
Rotation period:	17h 14m
Axial tilt:	97.8°

Uranian weather

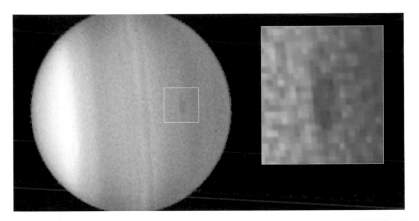

The unusual tilt of Uranus means that different parts of the planet experience different amounts of daylight. Some areas, for example, have a day and night corresponding to the planet's 17-hour rotation; others suffer decades of night followed by a brief period of 'daily' sunrise and sunset, and then more decades of permanent daylight. As the atmosphere adjusts its circulation to distribute heat around the planet, Uranus goes through a complex cycle of weather patterns. When *Voyager 2* flew past in 1986, the south pole was pointing almost directly at the Sun, and the flow of heat from pole to pole disrupted any weather systems that might have attempted to form in bands around the equator, producing a bland, featureless planet. Ten years later, the Hubble Space Telescope imaged a different planet, where spring had eased the polar circulation, allowing storms to form at last.

Feature of:	Uranus
Type of object:	Weather patterns including dark spot
Dark spot diameter:	3,000 x 1,700 km (1,900 x 1,100 miles)
Bright band width:	2,500km (1600 miles)
Date of image:	2006 (approaching 'equinox' for Uranus)

Uranus's rings

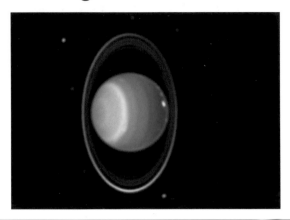

The system of rings around Uranus was discovered by chance in 1977, when astronomers studying the planet's occultation of a distant star noticed that the star flickered several times before and after the main eclipse, indicating that the light was dimmed by intervening rings. However, these rings are very different from those around Saturn; they are narrow bands of fairly dark material, probably frozen methane, which is less reflective than water ice. *Voyager 2* discovered a total of 11 rings, all just a few kilometres wide and deep. It is possible that the rings are kept confined by the gravity of shepherd moons close to Uranus, but they may be the remains of a ring system that was once far grander. According to this theory, ring systems are short-lived phenomena, created by the break-up of a moon or comet, and soon eroded by collisions and the gravity of their parent planet.

Main ring components:	Zeta Ring, Alpha Ring, Beta Ring, Delta Ring, Lambda Ring, Epsilon Ring, Nu Ring, Mu Ring
Overall radius:	38,000–97,730km (23,600–60,700 miles)
Typical thickness	About 100m (328m)
Size of particles	10cm > 10m (3.9in > 39in)
Composition	Methane ice?

Miranda

The innermost of Uranus's five major moons is the smallest, and was the last to be discovered, in 1948. Just 480km (298 miles) across, Miranda was found by Dutch–American astronomer Gerard Kuiper. In 1986, *Voyager 2* sent back surprising images. Rather than a cold, dead world, with a heavily cratered surface, they showed a Frankenstein moon, a patchwork of different terrains, jammed together at random and separated in many places by towering cliffs. Most impressive are the 'racetrack' features known as chevrons. Initially, Miranda was thought to have been the victim of a collision that shattered it into pieces, which later reassembled themselves. However, new evidence suggests that tidal heating from Uranus caused Miranda to partially melt and reshape itself.

Type of object:	Satellite of Uranus
Closest to Uranus:	129,222km (80,262 miles)
Furthest from Uranus:	129,558km (80,471 miles)
Orbital period:	1.41 days
Diameter:	472km (293 miles)
Rotation period:	1.41 days
Axial tilt:	Zero

Ariel

Uranus's second major moon, Ariel, is larger than Miranda, and considerably further away from its parent planet. It is a fairy typical rock/ice moon, with cratered plains separated by deep canyons, some of which have bright, icy floors. However, Ariel's craters are not as heavy or as large as on comparable worlds (such as Umbriel, the next moon out), which suggests that the moon was resurfaced and wiped clean at some point after its formation. It seems that Ariel is close enough to Uranus to have suffered from some tidal heating early in its history, which warmed the moon enough to trigger eruptions of icy 'cryovolcanic' slush similar to those that helped to shape some of Saturn's moons. The canyons were probably created by Ariel expanding as it cooled, allowing slush to well up again from within.

Type of object:	Satellite of Uranus
Closest to Uranus:	190,791km (118,503.6 miles)
Furthest from Uranus:	191,249km (118,788 miles)
Orbital period:	2.52 days
Diameter:	1157.8km (719.1 miles)
Rotation period:	2.52 days
Axial tilt:	Zero

Umbriel

The third major moon of Uranus is still mysterious, since *Voyager 2*, the only probe to have visited the Uranian system, flew close to Miranda but kept away from the outer moons. The handful of images reveal a dark, heavily cratered world. Umbriel seems to be caught between extremes: too far from its parent planet to have suffered much tidal heating, it is too small to have generated much internal heat of its own during formation. As a result, it remains largely unchanged since its formation, and like Jupiter's giant Callisto, acts as a target for any passing comet or asteroid that crosses Uranus's path. One intriguing feature is the bright 'splash' on Umbriel's upper limb. Despite first impressions, this is not a polar cap since *Voyager 2* photographed Umbriel from above; the splash actually lies close to the equator.

Type of object:	Satellite of Uranus
Closest to Uranus:	264,963km (164,573 miles)
Furthest from Uranus:	267,037km (165,861 miles)
Orbital period:	4.14 days
Diameter:	1169.4km (726.3 miles)
Rotation period:	4.14 days
Axial tilt:	Zero

85

Titania

Uranus's fourth major moon is its largest and, like Ariel, is dominated by cratered plains of various ages, and a number of canyons. Titania is too far from Uranus to have been heated by tidal forces, but seems to have been large enough to generate its own internal heat, powering icy volcanic activity. The moon's internal composition may have encouraged this; measurements of Titania's gravity suggest it is denser and rockier than many moons in the outer solar system. A rockier composition would contain more radioactive elements, whose slow decay would keep the interior warmer for longer. However, Titania's craters are still surrounded by fresh ice ejecta. While other moons are named after classical mythology, Uranus's satellites take their names from Shakespeare and Alexander Pope.

Type of object:	Satellite of Uranus
Closest to Uranus:	435,431km (270,454 miles)
Furthest from Uranus:	436,390km (271,050 miles)
Orbital period:	8.71 days
Diameter:	1577.8km (980 miles)
Rotation period:	8.71 days
Axial tilt:	Zero

Oberon

The outermost of Uranus's major moons is slightly smaller than Titania, and has a composition that is similarly rich in rocks. As a result, heat escaping from Oberon's interior was able to power icy volcanoes early in the moon's history, wiping away many early craters. However, Oberon froze solid more quickly than its sibling, and so it has a generally darker and more cratered surface. Large craters on Oberon offer a glimpse of its internal structure; their centres are marked by patches of darker ice, perhaps coloured by carbon compounds that have welled up from the depths after an impact has punctured the outer crust. One of Oberon's great mysteries is how it managed to form the mountain, over 6km (3.7 miles) high, seen on the limb in this picture.

Type of object:	Satellite of Uranus
Closest to Uranus:	582,703km (361,927 miles)
Furthest from Uranus:	584,337km (362,942 miles)
Orbital period:	13.46 days
Diameter:	1522.8km (945.8 miles)
Rotation period:	13.46 days (presumed)
Axial tilt:	Zero

87

Neptune

Neptune, the outermost major planet, was discovered in 1846 after French mathematician Urbain Le Verrier predicted its existence from changes in the orbit of Uranus. It is another ice giant, slightly smaller than Uranus and bluer in colour – a result of more methane in its atmosphere. Its axial tilt is roughly the same as Earth's, so it experiences normal seasons rather than the extremes of Uranus, although this far out in the solar system, the average temperature of its cloudtops is -200°C (- 320°F). This is about the same temperature as on Uranus, suggesting that Neptune has a strong internal heat source – probably powered by friction and chemical changes deep inside the planet. It is heat from the interior, not radiation from the Sun, that drives Neptune's powerful weather systems and high winds.

Type of object:	Ice giant planet
Closest to Sun:	4452 million km (2765.2 million miles)
Furthest from Sun:	4553 million km (2828 million miles)
Orbital period:	165.2 years
Diameter:	49,528km (30,762.7 miles)
Rotation period:	16h 7m
Axial tilt:	28.3°

The Great Dark Spot

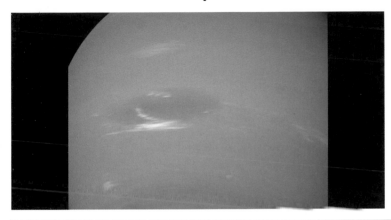

When *Voyager 2* flew past Neptune in 1989, it sent back views of an active world wracked by high winds and storms. The largest feature was a huge storm, named the Great Dark Spot (GDS) after its similarities to Jupiter's Great Red Spot. It, too, was a swirl of clouds high in the atmosphere, rising above a low-pressure area and allowing chemicals to flow up from below and condense as dark clouds in the upper atmosphere. Around the GDS sped high-altitude white clouds called 'scooters', carried by winds of up to 2200km/h (1350mph). The main difference between the Great Spots of Jupiter and Neptune seems to be their lifespans: while the Red Spot of Jupiter has persisted for centuries, studies of Neptune from Earth-based telescopes have shown that the GDS has now gone, replaced by other storms.

Feature of:	Neptune
Type:	Low-pressure atmospheric storm
Size:	c.12,000 x 8000km (7500 x 5000 miles)
Typical wind speeds:	2400km/h (1500mph)
Age:	Unknown

Triton

Compared to the other giant planets, Neptune's system of moons is strangely empty. A few satellites stay close to the planet, hugging its insubstantial ring system, while other minor worlds follow elliptical orbits further out. In between sits a single major moon, Triton, following an orbit that, though perfectly circular, goes the wrong way around the planet. Triton is an interloper from the Kuiper Belt, and its arrival wreaked havoc, sending the original moons out of Neptune's grasp, or at least into long elliptical orbits. *Voyager 2* revealed Triton to be an icy world of blues and browns, coated in frosts of chemicals. But its most impressive discovery was that Triton was an active, evolving world – dark streaks on the surface proved to be a result of geysers belching dust and liquid nitrogen into the tenuous atmosphere.

Type of object:	Satellite of Neptune
Closest to Neptune:	354,800km (220,373 miles)
Furthest from Neptune:	354,800km (220,373 miles)
Orbital period:	5.88 days (retrograde)
Diameter:	2706.8km (1681.2 miles)
Rotation period:	5.88 days (retrograde)
Axial tilt:	Zero

Canteloupe terrain

Triton's landscape is a mixture of two terrains: flat plains spotted with ice geysers, and a dimpled icescape known as canteloupe terrain after the pitted skin of the fruit. Together, they offer clues to Triton's past, explaining why it is still an active world, despite surface temperatures of around -237°C (-395°F). As Triton swung into its present orbit, it experienced powerful tides that slowed its rotation and heated its interior enough to turn much of the captured world into slushy chemical ice, with rafts of more solid material floating on top. Less dense ice rose through the molten parts of the surface in bubbles that dispersed to form the canteloupe terrain, while warm material trapped under the rafts created reservoirs of superheated nitrogen that escaped through weak points in the surface to form geysers.

Feature of:	Triton (satellite of Neptune)
Type:	Landscape with multiple depressions
Typical diameter of depressions:	30–50km (19–32 miles)
Composition:	Impure water ice
Age:	500 million years?

Nereid

Only the second moon of Neptune to be discovered, Nereid was found by Dutch-American astronomer Gerard Kuiper in 1949. Its orbit is one of the most elliptical in the solar system, ranging from 817,000km (507,660 miles) above Neptune at its closest approach, to 9.5 million km (5.9 million miles) at its most distant. Many astronomers believe that it is one of Neptune's original moons, ejected from its natural place by the capture of Triton. According to this theory, Nereid is a survivor among Neptune's original outer moons; the only satellites to survive the trauma almost unscathed were inner shepherd moons such as Proteus. *Voyager 2*'s only photograph of the moon, taken from a distance of 4.7 million km (2.9 million miles), showed it as a misshapen blob with a dark surface and hints of craters.

Type of object:	Satellite of Neptune
Closest to Neptune:	1,353,600km (840,745 miles)
Furthest from Neptune:	9,623,700km (5,977,453.4 miles)
Orbital period:	360.14 days
Diameter:	340km (211 miles)
Rotation period:	11h 31m
Axial tilt:	Unknown

Proteus

One of several new moons discovered close to Neptune during *Voyager 2*'s brief flyby in 1989, Proteus was found in time to redirect the cameras towards it. As a result, it is the only one of Neptune's smaller moons to have been photographed close up, revealing an ovoid world with a dark, heavily cratered surface dominated by a crater 255km (158 miles) across. Proteus is a shepherd moon of the type that accompanies and sustains the ring systems around other giant planets. Neptune has just three narrow ringlets, which are thicker on one side than the other – so thin in places that when astronomers first searched for rings around Neptune using the occultation method that had worked so well for Uranus, they believed they had found isolated 'ring arcs' that did not stretch all the way around the planet.

Type of object:	Satellite of Neptune
Closest to Neptune:	117,584km (73,034 miles)
Furthest from Neptune:	117,709km (73,111 miles)
Orbital period:	1.12 days
Size:	440 x 416 x 404km (273 x 258 x 251 miles)
Rotation period:	1.12 days
Axial tilt:	Zero

Pluto

Tiny Pluto is the brightest and among the largest members of the Kuiper Belt, a band of 'ice dwarf' worlds orbiting beyond Neptune. Discovered long before any of its cousins, it was for a long time considered a planet in its own right, and was demoted to 'dwarf planet' only in 2006, following the discovery of Eris*. Pluto was found by astronomer Clyde Tombaugh in 1930 during a deliberate search for a hypothetical (and now discredited) planet whose gravity was thought to be affecting Uranus and Neptune. For a long time, we knew very little about Pluto, but *Voyager 2*'s flyby of Triton in 1989 offered us a suggestion of a probable ice dwarf world, and not long after, the Hubble Space Telescope produced the first rough maps of its surface. It now seems that Pluto is a grey and brown world of rock and ice, covered in frosts of frozen methane and nitrogen.

Type of object:	Kuiper Belt Object/dwarf planet
Closest to Sun:	4436 million km (2755.3 million miles)
Furthest from Sun:	7376 million km (4581.4 million miles)
Orbital period:	248.1 years
Diameter:	2390km (1485 miles)
Rotation period:	6.4 days
Axial tilt:	119.6°

Charon

Pluto's giant moon Charon was discovered in 1978, when astronomer James Christy spotted a distortion in the distant world's tiny blob of light. It is, in fact, the largest moon in the solar system, relative to the size of its parent body, and it revealed that Pluto, like Uranus, is a world tipped on its side, experiencing decades-long seasons as it crawls round its orbit. What's more, because Charon orbits so close to Pluto, tidal forces have slowed the rotation of both worlds so that each now keeps one face permanently turned towards the other, and the two worlds rotate in the same period that Charon takes to orbit Pluto – 6.4 days. Charon is a similar world to Pluto, but more icy and red. In 2005, the Hubble Space Telescope discovered that it is actually the largest of three satellites in orbit around Pluto.

Type of object:	Satellite of Pluto
Closest to Pluto:	19,571km (12,156 miles)
Furthest from Pluto:	19,571km (12,156 miles)
Orbital period:	6.4 days
Diameter:	1207km (750 miles)
Rotation period:	6.4 days
Axial tilt:	Zero

1992 QB1

In 1951, the astronomer Gerard Kuiper suggested that Pluto, then seen as a planet at the edge of the solar system, might in fact be the largest member of a belt of similar worlds orbiting beyond Neptune, coinciding with the most distant points in the orbits of many short-period comets. The theory was proven when telescope technology became powerful enough, in the 1990s, to capture the faint light of these objects. 1992 QB1 was the first 'Kuiper Belt Object' to be found, using the Hubble Space Telescope and as the result of a deliberate search. At first indistinguishable from a faint star, it was identified as a slow-moving world beyond Neptune by its drift across a series of images taken several nights apart. 1992 QB1 is so small that little is known about it beyond an estimated diameter of 160km (99 miles).

Type of object:	Kuiper Belt Object
Closest to Sun:	40.9AU
Furthest from Sun:	46.6AU
Orbital period:	289.2 years
Diameter:	160km (99 miles)
Rotation period:	Unknown
Axial tilt:	Unknown

Quaoar

When found in 2002, Quaoar marked a breakthrough in the discovery of the Kuiper Belt – with a diameter just over half that of Pluto, it was a step up in size from other known Kuiper Belt worlds, and put an end to any lingering thoughts that Pluto might be special. Quaoar is named after a Native American creation god. It has a more circular orbit than most large Kuiper Belt Objects, orbiting at a more-or-less constant distance from the Sun of 43 astronomical units. It is also darker and redder than Pluto, Charon or Eris, which suggests that if it ever had a surface layer of ice, this has long since disappeared. Nonetheless, there is evidence for cryovolcanism or meteor impacts producing patches of ice on the surface. Quaoar has a recently discovered satellite with a diameter of roughly 100km (62 miles).

Type of object:	Kuiper Belt Object
Closest to Sun:	41.9AU
Furthest from Sun:	44.9AU
Orbital period:	286.0 years
Diameter:	1260km (783 miles)
Rotation period:	Unknown
Axial tilt:	Unknown

Eris

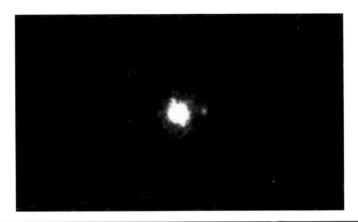

When discovered in 2003, Eris confused astronomers. It was the first Kuiper Belt Object to be found that was larger than Pluto, and forced astronomers to address Pluto's dual classification as both planet and Kuiper Belt Object. They eventually decided that rather than describe Eris (known at the time as 'Xena') as an official planet, they would downgrade Pluto to the new class of 'dwarf planets'. With its classification settled, the new world could be given an official name, and fittingly it was called Eris after the Greek goddess of discord. This new world, just 100km (62 miles) larger than Pluto, has a bright, frosted surface that makes it one of the most reflective objects in the solar system. Eris also has company on its long journey around the Sun – a moon called Dysnomia, just 150km (93 miles) across.

Type of object:	Kuiper Belt Object/dwarf planet
Closest to Sun:	5650 million km (3509.3 million miles) (37.8AU)
Furthest from Sun:	14,600 million km (9068.3 million miles) (97.6AU)
Orbital period:	557 years
Diameter:	2400km (1490.7 miles)
Rotation period:	8h?
Axial tilt:	Unknown

Sedna

Currently the most distant known member of the solar system, Sedna was discovered in 2003. It orbits the Sun in about 10,500 years, following an ellipse that clips the theoretical outer edge of the Kuiper Belt at its closest to the Sun, and goes out to almost 13 times that distance at its most remote. It therefore spends most of its time in a region of space astronomers believed to be empty, between the Kuiper Belt and the Oort Cloud. Appropriately for a world orbiting in the cold wastes of the solar system, Sedna is named after an Inuit goddess of the Arctic Ocean. Some astronomers believe it is the first member of an 'inner Oort cloud'; others believe it is an ice dwarf or asteroid ejected from closer to the Sun. Another mystery is Sedna's colour – it is among the reddest objects in the solar system.

Type of object:	Oort Cloud Object
Closest to Sun:	76.2AU
Furthest from Sun:	975.6AU
Orbital period:	About 12,059 years
Diameter:	About 1500km (930 miles)
Rotation period:	10h?
Axial tilt:	Unknown

Comet Halley

This comet owes its fame to the fact that it has been recorded during many of its returns to the inner Solar System since 240BC. Edmond Halley was the first to work out its 76-year orbit, in 1696, and we now know that it is among the youngest of the short-period comets that orbit in less than a couple of centuries. It reaches the outer edge of its orbit in the Kuiper Belt beyond Neptune, and its youth means that it retains large quantities of ice. During each passage round the Sun, some of this ice evaporates, forming an extended atmosphere called a coma, and often a spectacular tail stretching for millions of miles. Halley's predictability and brightness made it an obvious target for the first wave of comet probes, including the European *Giotto* mission, which had a close encounter with the comet's solid nucleus in March 1986.

Type of object:	Short-period comet
Closest to Sun:	0.6AU
Furthest from Sun:	35.1AU
Orbital period:	75.3 years
Size:	16 x 8 x 8km (10 x 5 x 5 miles)
Rotation period:	53h?
Axial tilt:	Unknown

Comet Hale-Bopp

The most impressive comet of recent times, Comet Hale-Bopp was discovered independently by two amateur astronomers, Alan Hale and Thomas Bopp, in 1995. At the time, it was more than a billion kilometers from the Sun, between the orbits of Jupiter and Saturn. Hale-Bopp, like most spectacular comets, was a rare visitor from the Oort Cloud, following an elliptical orbit lasting thousands of years. As it passed Jupiter in 1996, the giant planet's gravity deflected it into a new, considerably shorter orbit. By the time the comet rounded the Sun in April 1997, it outshone every star apart from Sirius, and was followed by gas and dust tails stretching up to 40° around the sky – an impressive sight despite the fact that it never came closer to Earth than the Sun.

Type of object:	Long-period comet
Closest to Sun:	0.9AU
Furthest from Sun:	371AU
Orbital period:	2537 years
Diameter:	About 40km (25 miles)
Rotation period:	11h 46m
Axial tilt:	Unknown

Comet Borrely

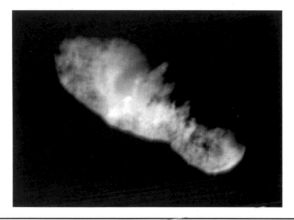

Comets that stray too close to Jupiter often have their orbits deflected by the giant planet's gravity, either flinging them out of the solar system altogether, or pulling them into orbits that take just a few years to go round the Sun. Comet Borrely fell victim to Jupiter at some point in its past, and since then its frequent returns to the Sun have boiled away most of its ice, leaving it as a dark and almost exhausted comet. When NASA's experimental *Deep Space 1* spacecraft flew past the comet in 2001, it sent back pictures of an elongated nucleus with a very faint coma of gas around it. The comet's surface has eroded away to create two distinct bulges joined by a narrow 'neck' – a future passage around the Sun or close to Jupiter could see the comet broken apart entirely by the forces of heat and gravity.

Type of object:	Short-period comet
Closest to Sun:	1.4AU
Furthest from Sun:	5.8AU
Orbital period:	6.8 years
Size:	8 x 4 x 4km (5 x 3 x 3 miles)
Rotation period:	26h?
Axial tilt:	Unknown

Comet Wild 2

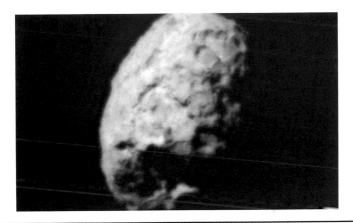

This young comet entered the inner solar system for the first time quite recently, and after a close encounter with Jupiter its orbit was shortened to just 6.4 years. This made it an ideal target for NASA's *Stardust* mission, an ambitious attempt to fly behind a comet and collect particles from its tail. *Stardust* caught up with the comet in 2004, and returned its delicate cargo of particles to Earth in a re-entry capsule in 2006. Comet tails are created as the solar wind interacts with the particles that burst from beneath a comet's dark and dusty crust when it heats up closer to the Sun. Frequently a comet will develop two distinct tails – a bluish-white ion tail of glowing gases that always points directly away from the Sun, and a yellowish dust tail of heavier particles that curves back along the comet's path through space.

Type of object:	Short-period comet
Closest to Sun:	1.6AU
Furthest from Sun:	5.3AU
Orbital period:	6.4 years
Diameter:	5.5km (3.4 miles)
Rotation period:	12h?
Axial tilt:	Unknown

Tempel 1

In 2005, NASA launched *Deep Impact*, its most ambitious comet exploration probe yet. This robot spacecraft was designed to orbit around comet Tempel 1, and then fire a 370kg (816lb) projectile into its surface at high speed, creating an artificial crater that would fling out material from beneath the comet's surface for the orbiter to analyze. The impact occurred on 4 July 2005, and *Deep Impact* took this picture moments later. Unfortunately the brightness of the dust cloud that erupted from the impact site thwarted attempts to observe the crater. However, studies of the ejecta material brought a few surprises – the comet proved to have more dust and less ice than expected, with a rich variety of minerals. Astronomers had also expected the dust to have sandlike grains, but the ejecta turned out to be a fine powder.

Type of object:	Short-period comet
Closest to Sun:	1.5AU
Furthest from Sun:	4.7AU
Orbital period:	5.5 years
Size:	7.6 x 4.9km (4.7 x 3.0 miles)
Rotation period:	41h
Axial tilt:	Unknown

Oort Cloud

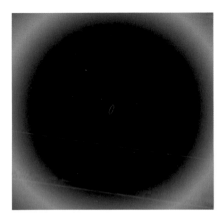

Proposed by Dutch astronomer Jan Oort in 1950 (inspired by the Estonian Ernst Öpik), the Oort Cloud cannot be seen but almost certainly exists, and marks the outer edge of our solar system. A spherical shell of dormant comets, it lies between half and one light year from the Sun, at the limit of its gravitational grasp. Oort proposed it to explain why so many long-period comets reach their greatest distance from the Sun in this region of space. The cloud may contain trillions of comet nuclei, which fall towards the inner solar system only when disturbed by, for instance, a collision or the gravitational tug of a passing star. The Oort Cloud comets are thought to have formed closer to the Sun, and been ejected into the cold wastes of space by encounters with the giant planets shortly after their creation.

Object type:	Comet cloud
Structure:	Hollow spherical shell extending between 0.5 and 1 light year From the Sun. Possible inner extension towards Kuiper Belt.
Content:	Several million million comet nuclei?
Mass:	About 3 Earth masses
Age:	4.6 billion years

Northern circumpolar stars

Across most of the northern hemisphere, the stars on this map are permanent fixtures of the northern sky, changing only their orientation as they pivot through the course of a day around the north celestial pole. Three constellations dominate this area of the sky: Ursa Minor, Cassiopeia and Ursa Major. Ursa Major, the Great Bear, is the best-known, thanks to the Plough or Big Dipper pattern formed by its seven brightest stars. The Little Bear Ursa Minor, whose tail-star, Polaris, marks the celestial pole almost exactly, has roughly the same shape as the Plough, while Cassiopeia has a distinctive W-shape and counterbalances Ursa Major on the opposite side of Polaris. Following a line through Dubhe and Merak, the stars on the side of the 'dipper' lead directly to Polaris, while the curve of its handle, extended across the sky, leads first to Arcturus in Bootes, and then to Spica in Virgo.

Centre of projection:	RA: 0h 00m, Dec: 90°
Key constellations:	Camelopardalis, Cassiopeia, Cepheus, Draco, Lacerta, Lynx, Perseus, Ursa Major, Ursa Minor
Opposite the Sun:	N/A
Visibility:	All year round for latitudes above 40° N

Winter stars

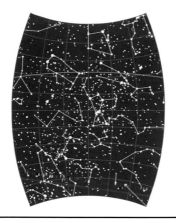

The constellations shown on this map lie opposite the Sun around December and January, when most are visible through the night for most of the northern and southern hemispheres. These constellations of the northern winter are dominated by Orion, the hunter, at the centre of a tableau that also involves the bull Taurus and the dogs Canis Major and Minor. Further to the north lie the constellations of Auriga, the Charioteer, and Gemini, the Twins, while to the south lie the long celestial river Eridanus and Puppis, the stern, part of the great ship Argo Navis. Orion's stars offer several 'pointer' alignments. A line through the three stars of the belt points roughly towards Sirius to the southwest, and Aldebaran to the northeast. Rigel at Orions foot forms a near-equilateral triangle with Sirius and Procyon, while a diagonal from Rigel through Betelgeuse points at Castor and Pollux in Gemini.

Centre of projection:	RA: 6h 00m, Dec: 0°
Key constellations:	Auriga, Cancer, Canis Major, Canis Minor, Caelum, Columba, Eridanus, Gemini, Lepus, Monoceros, Orion, Perseus, Puppis, Taurus
Opposite the Sun:	Mid-December
Visibility:	Northern winter, southern summer

Autumn stars

The skies of northern autumn and southern spring are comparatively barren. For most observers, the standout feature is the Square of Pegasus, a large rectangle of bright stars with an almost empty central region. Rooted to the northwestern corner of the Square is the branching shape of Andromeda, with the nearby spiral galaxy M31 (the Andromeda Galaxy) easily visible to the naked eye. Around the celestial equator, a watery group of constellations includes Pisces (the Fish) and Aquarius (the Water Carrier), the sea monster Cetus (home to Mira, one of the sky's finest variable stars), and the southern fish Piscis Austrinus, with its prominent bright star Fomalhaut. Most of the constellations shown on this chart are visible throughout the night for observers at middle latitudes between the months of September and October, when they lie roughly opposite the Sun in the sky.

Centre of projection:	RA: 0h 00m, Dec: 0°
Key constellations:	Andromeda, Aquarius, Aries, Cetus, Equuleus, Fornax, Pegasus, Pisces, Piscis Austrinus, Sculptor, Triangulum
Opposite the Sun:	Mid-September
Visibility:	Northern autumn, southern spring

Spring stars

During northern spring (southern autumn), the skies are are dominated by zodiac constellations, including Leo, the Lion, and Virgo, the Maiden. Leo's brightest star, Regulus, lies almost directly on the ecliptic, and often interacts with the Moon and planets, in events such as conjunctions (close encounters) and occultations (in which the star is actually eclipsed, usually by the Moon). Spica in Virgo is slightly further from the ecliptic, and a line drawn between the two bright stars more or less follows the ecliptic. South of this line lies the sinuous train of stars that make up the Water Snake Hydra, the sky's largest constellation. More compact nearby groups, such as Corvus and Crater, can be easier to spot. On the other side of the ecliptic, Leo, Virgo and Coma Berenices offer unparalleled views of relatively nearby galaxies, including the closest galaxy cluster to our own.

Centre of projection:	RA: 12h 00m, Dec: 0°
Key constellations:	Antlia, Bootes, Canes Venatici, Centaurus, Coma Berenices, Corvus, Crater, Hydra, Leo, Leo Minor, Sextans, Vela, Virgo
Opposite the Sun:	Mid-March
Visibility:	Northern spring, southern autumn

Summer stars

The skies of the northern summer and southern winter are the richest of all, overwhelmed by a bright swathe of stars scattered around the Milky Way. These stars are at their best for mid-latitude observers around June and July, when they lie opposite the Sun in the sky and are visible through the night. At one corner of this swathe of sky, the cross-shape of Cygnus, the Swan, flies down the northern Milky Way. Its brightest star, Deneb, forms one corner of the northern 'Summer Triangle' with brilliant Vega in Lyra (the Lyre), and Altair in Aquila (the Eagle), near the celestial equator. The Milky Way sweeps south from here to Sagittarius the Archer, where it is at its most dense, studded with bright star clusters and emission nebulae that mark the direction of our galaxy's centre. Other constellations, such as Scorpius and parts of Serpens and Ophiuchus, are also rich in similar features.

Centre of projection:	RA: 18h 00m, Dec: 0°
Key constellations:	Aquila, Capricornus, Corona Australis, Corona Borealis, Cygnus, Delphinus, Hercules, Libra, Lupus, Lyra, Ophiuchus, Sagitta, Sagittarius, Scorpius, Scutum, Serpens, Vulpecula
Opposite the Sun:	Mid-June
Visibility:	Northern summer, southern winter

Southern circumpolar stars

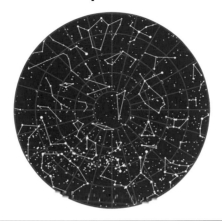

The southern circumpolar sky has two distinct halves. One side is crossed by a bright swathe of the southern Milky Way and prominent constellations such as Centaurus, Crux (the Southern Cross), and Vela, Carina and Puppis (the Sail, Keel and Stern of the great ship Argo). The area includes some of the brightest stars in the sky, including the nearby Alpha Centauri system, and the more distant Canopus. Crux is a useful pointer towards the south celestial pole which, unlike its northern counterpart, has no bright star to mark it. On the opposite side of the pole, and in the regions directly around it, the skies are more barren, with a jumble of small, faint constellations. Achernar at the end of Eridanus is the only bright star in the region, although other highlights are the Magellanic Clouds in Tucana and Dorado.

Centre of projection:	RA: 0h 00m, Dec: -90°
Key constellations:	Ara, Carina, Centaurus, Chamaeleon, Circinus, Crux, Dorado, Eridanus, Grus, Horologium, Hydrus, Indus, Lupus, Mensa, Musca, Norma, Octans, Pavo, Phoenix, Puppis, Pictor, Reticulum, Telescopium, Triangulum Australe, Tucana, Vela, Volans
Opposite the Sun:	N/A
Visibility:	All year round for latitudes below 40° S

Hydra

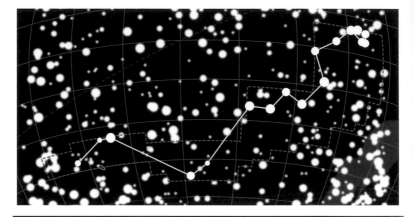

The great constellation of Hydra runs along a stretch of the celestial equator, through a fairly barren stretch of sky. It represents the water snake that features in the legend of Corvus and Crater*, and despite its size it has only one really noticeable star. Magnitude 2.0 Alphard, whose name means 'the solitary one', is an orange giant star thought to have an invisible white dwarf companion – the remains of a more massive star that has long since aged and died, flinging off its outer layers and enriching the atmosphere of its neighbour with unusual elements such as barium. An attractive open star cluster, M48, lies south of the distinctive cluster of stars marking Hydra's head – it is just visible with the naked eye, and makes a good target for binoculars. M83, in Hydra's tail, is the Southern Pinwheel, a face-on spiral galaxy some 15 million light years away, visible through a small telescope.

Latin name:	Hydra
Latin genitive:	Hydrae
Abbreviation:	Hya
Meaning:	Water Snake
Visibility:	Northern spring, southern autumn

Virgo

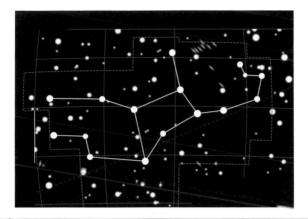

This constellation has represented a celestial maiden since the earliest records – in Classical times, it was usually associated with Persephone, daughter of the harvest goddess Ceres. The brightest star, Spica*, represents an ear of wheat held in the maiden's hand, and Virgo's association with the harvest also goes a long way back (since around the time the original constellations developed, the Sun was in Virgo at harvest time). Virgo is easy to find in the northern hemisphere by extending an arc from the curving handle of the Plough in Ursa Major, through Arcturus in Boötes, until it intersects with Spica. Gamma (γ) Virginis, or Porrimma, is a beautiful pair of twin white stars in a binary system, currently divisible for moderate telescopes, and getting easier as its stars draw apart in their orbits. However, Virgo's chief attraction is the Virgo Cluster – the closest major galaxy cluster to the solar system*.

Latin name:	Virgo
Latin genitive:	Virginis
Abbreviation:	Vir
Meaning:	Maiden
Visibility:	Northern spring, southern autumn

Ursa Major

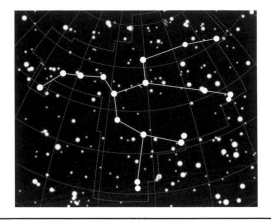

The seven brightest stars of the Great Bear form the most famous star pattern, or 'asterism', in the sky: the Plough or Big Dipper. The constellation as a whole represents Callisto, a beautiful woman who was transformed into a bear by the jealous Juno, queen of the gods, and it extends well beyond the central bright stars. Five of the Plough stars are moving in the same direction through space – they form the Ursa Major Moving Group, the nearest star cluster to our own. Mizar, the middle star in the Plough's handle, is a celebrated multiple star 78 light years away. With either the naked eye or binoculars, it is easy enough to spot its companion Alcor, which just happens to lie in the same direction in the sky, but a small telescope will also show that Mizar has a much closer binary companion – and in fact, each of these stars is double in its own right. Elsewhere, Ursa Major contains several beautiful galaxies*.

Latin name:	Ursa Major
Latin genitive:	Ursae Majoris
Abbreviation:	UMa
Meaning:	Great Bear
Visibility:	Northern hemisphere circumpolar

Cetus

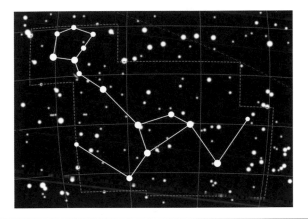

This large constellation to the southwest of Taurus represents a whale or sea monster, often called Typhon. It features in many legends of the sky, most notably in the tale of Perseus and Andromeda. Cetus is home to the famous variable star Omicron (o) Ceti or Mira*, which fades from brightness to invisibility and back in an 11-month cycle. Gamma (γ) Ceti is a close double star divisible in small telescopes, with stars of magnitudes 3.5 and 7.3, which sometimes appear bluish and yellowish. Elsewhere in the constellation lie Luyten 726-8*, one of the closest star systems to the Sun*, and Tau (τ) Ceti, a fairly sunlike star of magnitude 3.5 and just 12 light years away, surrounded by its own 'Kuiper Belt' of cool material. M77* is an attractive galaxy for a small telescope, with an unusually bright core – it is classified as a Seyfert Galaxy*.

Latin name:	Cetus
Latin genitive:	Ceti
Abbreviation:	Cet
Meaning:	Whale/Sea Monster
Visibility:	Northern autumn, southern spring

Hercules

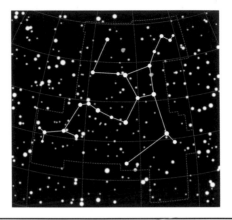

This sprawling constellation can be hard to locate thanks to its lack of really bright stars – its most identifiable feature is the lopsided rectangle known as the Keystone, which marks the body of the Greek hero, with chains of stars extending from each corner to form his limbs. The brightest star, of magnitude 3.5, is Rasalgethi, whose name means 'the kneeler's head' (because the constellation is actually 'upside down' for the northern hemisphere, with Hercules kneeling on top of the dragon Draco). Rasalgethi is a multiple star – small telescopes separate it into a red giant that varies erratically between magnitudes 3 and 4, and a whitish companion that is itself a double when viewed with larger instruments. The constellation's best deep-sky objects are the globular clusters M13* and M92. It also has a relatively bright planetary nebula, NGC 2610, which appears as a greenish disc in small telescopes.

Latin name:	Hercules
Latin genitive:	Herculis
Abbreviation:	Her
Meaning:	Hercules
Visibility:	Northern summer, southern winter

Eridanus

The long constellation of the celestial river winds all the way from the foot of Orion close to the celestial equator, to brilliant Achernar ('River's End') in the far southern skies. It is supposedly the river into which Phaeton, son of the Sun god Helios, crashed his father's chariot after stealing it. Achernar*, or Alpha (α) Eridani, is a hot blue-white star, 143 light years away yet brilliant at magnitude 0.5. Elsewhere in the constellation lie Epsilon (ε) Eridani*, a sunlike star with at least one planet, and the 40 Eridani system, which includes the sky's most easily seen white dwarf. NGC 1535, meanwhile, is on its way to forming another white dwarf – it is a compact planetary nebula called Cleopatra's Eye, 5800 light years away and visible through a small telescope. A medium-sized instrument will show its central star of magnitude 12.2.

Latin name:	Eridanus
Latin genitive:	Eridani
Abbreviation:	Eri
Meaning:	River
Visibility:	Northern winter, southern summer

Pegasus

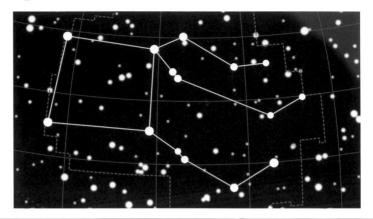

The famous constellation of the winged horse is dominated by the Square of Pegasus, a near-perfect box of bright stars. Chains of stars marking the horse's forelegs, neck and head emerge from the western stars of the Square, while the northeastern star, Alpheratz, is the root for the branching shape of Andromeda. In fact, Alpheratz is both Delta (δ) Pegasi and Alpha (α) Andromedae. The brightest star in the constellation, however, is Epsilon (ε) Pegasi or Enif, marking the horse's nose (the figure is upside down when viewed from the northern hemisphere). This orange supergiant, 670 light years away, normally shines at magnitude 2.4, but in 1972 it suddenly flared to magnitude 0.8 for just a few minutes – an event that has still not been fully explained. Elsewhere in the constellation lie 51 Pegasi*, one of the first extrasolar planetary systems, and M15, a beautiful globular cluster just below naked-eye visibility.

Latin name:	Pegasus
Latin genitive:	Pegasi
Abbreviation:	Peg
Meaning:	Winged Horse
Visibility:	Northern autumn, southern spring

Draco

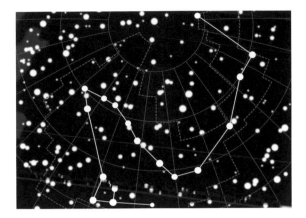

This constellation represents a dragon fought by Hercules as one of his 12 tasks, and its serpentine form wraps itself most of the way around Ursa Minor and the north celestial pole. Despite its size, it has few bright stars. Alpha (α) Draconis or Thuban is one of many 'Alpha' stars that are not actually the brightest in their constellations – it is a yellow-white giant about 310 light years away and shining at magnitude 3.5. Its chief claim to fame is that precession (the long, slow 'wobble' in Earth's axis of rotation) made Thuban the pole star in the time of the Egyptian pyramid builders. Mu (μ) Draconis, also known as Arrakis, is an attractive double star for small or medium-sized telescopes, while 39 Draconis is a complex multiple system of at least eight members (the brightest and widest can be separated with binoculars or a small telescope). NGC 6543 is the Cat's Eye*, a beautiful planetary nebula.

Latin name:	Draco
Latin genitive:	Draconis
Abbreviation:	Dra
Meaning:	Dragon
Visibility:	Northern hemisphere circumpolar

Centaurus

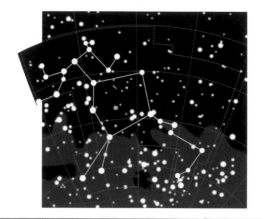

This large and bright constellation represents Chiron, the wise centaur that tutored both Hercules and Achilles in Greek mythology. In its southeast corner lies Alpha (α) Centauri*, the closest star system to the Sun, and nearby is Beta (β) Cen, or Hadar. Almost as bright as Alpha at magnitude 0.6, Hadar lies 525 light years away and is far more luminous. Medium-sized telescopes will split the star to reveal a magnitude 4 companion, but the brighter primary is also a close spectroscopic binary. Each star in the system is massive – the fainter companion has five times the mass of the Sun, while the brighter close binary's elements both weigh as much as 15 Suns. The central polygon of the centaur's body, meanwhile, contains two interesting deep-sky objects – the naked-eye globular cluster Omega Centauri* and the active galaxy Centaurus A*, both visible through binoculars on dark nights.

Latin name:	Centaurus
Latin genitive:	Centauri
Abbreviation:	Cen
Meaning:	Centaur
Visibility:	Northern spring, southern autumn

Aquarius

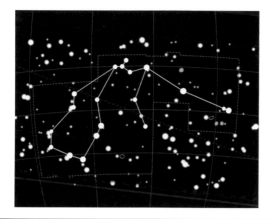

Although to modern eyes it appears fairly shapeless and relatively faint, Aquarius is one of the oldest constellations. The ancient Greeks saw the figure pouring water from a jug as Ganymede, a handsome youth abducted by Zeus and taken to serve the gods on Mount Olympus. Zeta (ζ) Aquarii is an attractive binary for small telescopes, consisting of near-twin stars of magnitudes 4.3 and 4.5, but the constellation's chief attractions are its deep-sky objects. M2 is a bright globular cluster, 37,500 light years away and easily spotted through binoculars or a small telescope. It is unusual for being noticeably elliptical rather than spherical in shape. NGC 7293 is the Helix Nebula*, the closest planetary nebula to Earth but, unfortunately, a faint and difficult object for amateur observers. NGC 7009 is the Saturn Nebula – more distant and compact but considerably easier to spot in a small telescope.

Latin name:	Aquarius
Latin genitive:	Aquarii
Abbreviation:	Aqr
Meaning:	Water Carrier
Visibility:	Northern autumn, southern spring

Ophiuchus

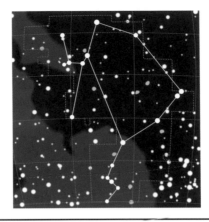

This large constellation has been seen as a man wrestling or carrying a serpent since the earliest times, even though its pattern is hardly distinctive to modern eyes. The serpent-bearer's identity has varied through time, from the heroes Jason and Cadmus to the more benevolent modern interpretation as Asclepius, god of medicine, with the snake wrapped around a magical staff. The southern reaches of Ophiuchus are crossed by the ecliptic (the path made by the Sun as it seems to move across the sky throughout the year), and so the Moon and planets can appear here too, making Ophiuchus the unofficial 13th sign of the zodiac. Rho (ρ) Ophiuchi is an attractive multiple star, with four well-spaced stars easily visible through binoculars. The system is still embedded in star-forming nebulosity, visible only through more powerful telescopes. Elsewhere lies Barnard's Star*, a faint red dwarf visible only through binoculars despite being the fourth closest star to the Sun.

Latin name:	Ophiuchus
Latin genitive:	Ophiuchi
Abbreviation:	Oph
Meaning:	Serpent Bearer
Visibility:	Northern summer, southern winter

Leo

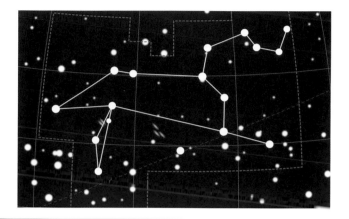

Of all the constellations, this one bears perhaps the strongest resemblance to the creature it represents – a recumbent or crouching lion. In Greek myth, it represented the Nemaean lion, fought by Hercules, but even peoples that had no exposure to classical culture recognized the figure of a big cat in this part of the sky. Alpha (α) Leonis is Regulus, a bright star which is of magnitude 1.4 and about 77 light years from the Sun. It produces 150 times the light of the Sun, and spins rapidly – once every 15 hours. R Leonis is a slowly pulsating red giant variable similar to Mira in Cetus. It lies 390 light years away, and is best tracked with binoculars or a small telescope – at its brightest, it is just visible to the naked eye; while at its faintest it is beyond the range of binoculars. Algieba (γ Leonis) is an attractive binary with orange and yellow components that are easily split in a small telescope.

Latin name:	Leo
Latin genitive:	Leonis
Abbreviation:	Leo
Meaning:	Lion
Visibility:	Northern spring, southern autumn

Boötes

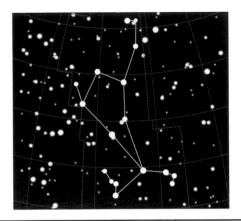

Boötes represents a herdsman driving the Great and Little Bears around the celestial pole with the aid of his hunting dogs, Canes Venatici. Its distinctive kite shape is easy to spot thanks to the brilliant orange Arcturus (magnitude 0.0) at its base. When the curve of the Plough's handle (the tail of the Great Bear) is extended in an arc, it points directly to Arcturus*, the giant star closest to the solar system. Elsewhere in the constellation lies the attractive double Epsilon (ε) Boötis or Izar. A decent telescope is needed to split this star into its contrasting components – an orange giant of magnitude 2.7 and a blue-white star of magnitude 5.1, which move in a 1000-year orbit around each other. Tau (τ) Boötis*, meanwhile, is one of the brightest stars known to have its own planetary system. It is a white star of magnitude 4.5, 51 light years away.

Latin name:	Boötes
Latin genitive:	Boötis
Abbreviation:	Boo
Meaning:	Herdsman
Visibility:	Northern spring, southern autumn

Pisces

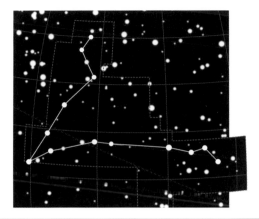

Two long straggling lines of stars meeting at the magnitude 3.8 Alrescha ('the cord'), Pisces is generally depicted as a pair of fish tied together at their tails. Originally, however, the fish were swimming free, and represented Aphrodite and her son Cupid fleeing from the approach of the sea monster Cetus. Alrescha or Alpha (α) Piscium is a binary pair of white stars that can be separated in a medium-sized telescope. Its stars, with magnitudes 4.2 and 5.2, complete an orbit of each other every 720 years, and the fainter star may also be a spectroscopic binary. Zeta (ζ) Piscium is a similar pairing of white stars with magnitudes 5.2 and 6.3. Although its stars are fainter, they are more widely spaced and therefore easier to resolve with smaller instruments. M74, meanwhile, is a beautiful face-on spiral galaxy, 30 million light years away and hard to spot because its light is spread across a wide area of sky.

Latin name:	Pisces
Latin genitive:	Piscium
Abbreviation:	Psc
Meaning:	Fish
Visibility:	Northern autumn, southern spring

Sagittarius

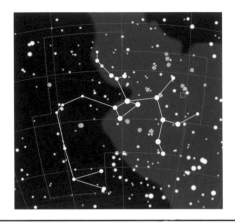

Although it is supposed to represent a warrior centaur armed with bow and arrow, the central region of Sagittarius most closely resembles a celestial teapot to modern eyes. The constellation happens to lie towards the centre of our galaxy, and as a result is rich in deep-sky objects and very rewarding to any observer. Epsilon (ε) Sagittarii, or Kaus Australis, is a bright white giant of magnitude 1.8, 145 light years away and mysteriously lacking in the heavy elements that should be common in such a star. Beta (β) Sagittarii is Arkab, an attractive 'line of sight' double, in which one star is also a physical binary. A chain of nebulae runs through the northwestern corner of the constellation, including the Lagoon, Trifid* and Omega Nebulas – a great hunting ground for binoculars or small telescopes. Elsewhere, radio emissions reveal the true heart of our galaxy – an object called Sgr A*.

Latin name:	Sagittarius
Latin genitive:	Sagittarii
Abbreviation:	Sgr
Meaning:	Archer
Visibility:	Northern summer, southern winter

Cygnus

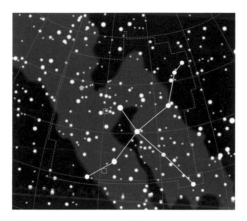

The distinctive cross-shape of Cygnus represents a swan flying down the rich starfields of the northern Milky Way. It is a constellation that has been blessed with a great range of interesting and beautiful objects. Its brightest star, Deneb*, is an impressive white supergiant, the most distant and truly luminous of the sky's bright stars. And P Cygni, an unpredictable variable star of about magnitude 4.8, is an even more awe-inspiring blue supergiant some 5000 light years away – one of the most luminous stars in the entire sky, it pumps out the energy of 700,000 suns. Beta (β) Cygni is Albireo*, a celebrated double star, while NGC 7000 is the extensive North America Nebula, a huge star-forming region that appears as a glowing patch of light through binoculars. Cygnus is also home to a black hole binary system, a bright supernova remnant, and a luminous radio galaxy.

Latin name:	Cygnus
Latin genitive:	Cygni
Abbreviation:	Cyg
Meaning:	Swan
Visibility:	Northern summer, southern winter

Taurus

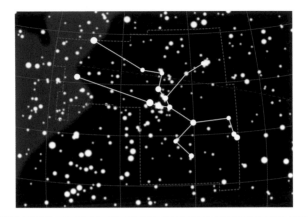

The constellation of the Bull is one of the few that truly resembles the creature it supposedly represents – a charging beast rearing to attack the great hunter Orion. It is easy to locate thanks to the brilliance of its bright star Aldebaran and the distinctive V-shape of the Hyades* cluster marking the bull's face. The constellation is also rich in interesting objects: aside from the Hyades, there is another even more beautiful and famous star cluster – the Pleiades, or Seven Sisters*. Close by the tip of the bull's southern horn lies the sky's finest supernova remnant – the Crab Nebula, which is just visible with binoculars and is easily found with a small telescope. Elsewhere lies the young variable star T Tauri*, still in the process of formation. Aldebaran itself is an orange giant. Although it lies in the same direction as the Hyades, it is in fact considerably closer to Earth, at a distance of 65 light years.

Latin name:	Taurus
Latin genitive:	Tauri
Abbreviation:	Tau
Meaning:	Bull
Visibility:	Northern winter, southern summer

Camelopardalis

This obscure constellation was introduced to the northern sky by Dutch astronomer and theologian Petrus Plancius in the early seventeenth century. It supposedly represents a beast of burden that carried Rebecca to her wedding with Isaac in the biblical story, but its Latin name actually means 'the Giraffe'. Beta (β) Cam is a multiple star – small telescopes will reveal that this yellow supergiant with a magnitude of 4.0 has a companion of magnitude 8.6, and larger instruments will show that this star is itself double. The entire system lies about 1000 light years away, and the yellow supergiant is something of a mystery: in 1967, it suddenly jumped in brightness by a whole magnitude for just a few minutes, perhaps due to an enormous stellar flare. NGC 2403, meanwhile, is an attractive face-on spiral best seen through good binoculars. It was first recorded in 1788 by William Herschel, discoverer of Uranus.

Latin name:	Camelopardalis
Latin genitive:	Camelopardalis
Abbreviation:	Cam
Meaning:	Giraffe
Visibility:	Northern hemisphere circumpolar

Andromeda

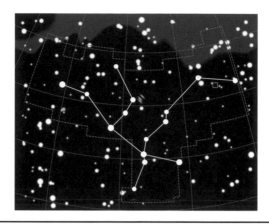

Easily found by tracking up from bright Alpheratz on the northeastern corner of the Square of Pegasus, Andromeda represents a princess from the mythical story of Perseus, chained to a rock and awaiting sacrifice to the sea monster Cetus. The constellation is most famous for the great Andromeda Galaxy M31* – a bright spiral that is the closest major galaxy to our own. At a distance of 2.9 million light years, M31 is the most distant object visible with the naked eye, and it has satellite galaxies that can be seen with small telescopes. Alpheratz is Alpha (α) Andromedae – a blue-white star of magnitude 2.1 that is prone to slight variations as it rotates. Gamma (γ) Andromedae, or Almach, is a beautiful triple star: small telescopes can split it to reveal a yellow star of magnitude 2.3 and a blue-white companion of magnitude 4.8, but at present only larger instruments can reveal the blue star's close companion of magnitude 6.1.

Latin name:	Andromeda
Latin genitive:	Andromedae
Abbreviation:	And
Meaning:	Andromeda
Visibility:	Northern autumn, southern spring

Puppis

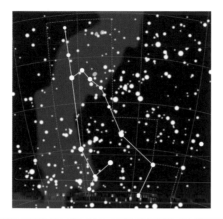

This bright constellation, embedded in the Milky Way to the south of Canis Major, represents the stern of the great ship Argo. Comprising present-day Puppis, Carina and Vela, it was the largest constellation in the sky until it was broken apart by Nicolas de Lacaille in the 1750s. Its brightest star, Zeta (ζ) Puppis or Naos, is a stellar monster with the mass of 60 suns. Lying 1400 light years from Earth, it pumps out 21,000 times as much visible light as the Sun, but because its surface is so hot (about 42,000°C/76,000°F), most of its energy must be released as ultraviolet radiation, suggesting a total energy output 750,000 times that of the Sun. NGC 2451 is a small open cluster of about 40 stars visible to the naked eye and dominated in binocular views by the orange giant c Puppis. M46 is a more distant but richer cluster for binoculars.

Latin name:	Puppis
Latin genitive:	Puppis
Abbreviation:	Pup
Meaning:	Stern
Visibility:	Northern winter, southern summer

Auriga

This constellation is easily found thanks to its chief star, Capella – the sixth brightest star in the sky at magnitude 0.1. Its other noticeable feature is a narrow triangle of stars known as 'The Kids'. This asterism has been associated with a group of goats for just as long as the overall constellation has been a Charioteer. Capella, or Alpha (α) Aurigae, is actually a quadruple star – a tightly bound spectroscopic binary with two yellow giants pumping out 50 and 80 Suns' worth of energy, orbited by a more distant pair of red dwarfs, which are visible through a medium-sized telescope. The southern stars of the Kids are known as Haedus I and II. Haedus I or Zeta (ζ) Aur is an eclipsing binary with an unusually long period: it dips in brightness by 0.15 magnitudes for 36 days in a 32-month cycle. Almaaz or Epsilon (ϵ) Aurigae* is another eclipsing binary, though a very strange one.

Latin name:	Auriga
Latin genitive:	Aurigae
Abbreviation:	Aur
Meaning:	Charioteer
Visibility:	Northern winter, southern summer

Aquila

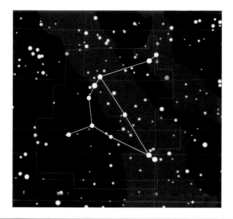

The ancient constellation of the Eagle is easily located thanks to its bright star Altair, the southern point of the northern sky's 'Summer Triangle' (completed by Deneb in Cygnus and Vega in Lyra). The pattern represents Zeus, king of the Greek gods, who transformed himself into an eagle in order to abduct Ganymede (nearby Aquarius). Altair is a nearby white star, 17 light years from Earth and 11 times as luminous as the Sun. It shines at magnitude 0.8 in our skies, flanked by a pair of slightly fainter stars – Alshain and Tarazed. Alshain is Beta (β) Aquilae, a yellow giant 45 light years from Earth, while Tarazed is Gamma (γ) Aql, an orange giant 460 light years away. Astronomers have found that Tarazed is unusually large for an orange giant – big enough to engulf Mercury and reach almost to Venus if placed in our own solar system.

Latin name:	Aquila
Latin genitive:	Aquilae
Abbreviation:	Aql
Meaning:	Eagle
Visibility:	Northern summer, southern winter

Serpens

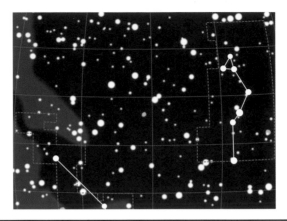

This constellation represents the snake with which Ophiuchus, the serpent-bearer, is wrestling, and is unique in the sky as the only constellation that is split into two parts. Serpens Caput, the snake's head, lies to the west of Ophiuchus, and Serpens Cauda, its tail, lies to the east, in a dense region of the Milky Way. Alpha (α) Serpentis is Unukalhai, the 'serpent's neck'. It is an orange giant 73 light years away, and is thought to be in the period of relative calm, when a star near the end of its life briefly finds respite by burning the helium residue in its core. M5 in Serpens Cauda is an impressive globular cluster 24,500 light years away, just visible with the naked eye and an impressive sight with binoculars, while M16 is the young open cluster at the heart of the famous Eagle Nebula*, a fertile region of starbirth.

Latin name:	Serpens
Latin genitive:	Serpentis
Abbreviation:	Ser
Meaning:	Serpent
Visibility:	Northern summer, southern winter

Perseus

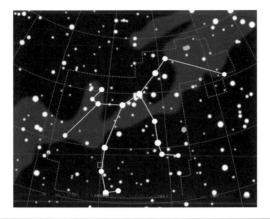

This bright but misshapen constellation represents Perseus, hero of a legend that encompasses figures from across the sky. Perseus was favoured by the goddess Athena, who equipped him to kill the gorgon Medusa. He then used Medusa's head to turn the sea monster Cetus to stone, rescuing the princess Andromeda, daughter of Cepheus and Cassiopeia. The brightest star, Alpha (α) Persei or Mirfak, is a white supergiant of magnitude 1.8, embedded in an open cluster called Melotte 20. The cluster is 50 million years old, 590 light years away, and a rewarding target for binoculars. Beta (β) Per is the famous eclipsing binary star Algol*. Algol means 'the demon', and it marks the baleful eye of Medusa. Perseus is rich in star clusters, and the finest is the Double Cluster, NGC 869/884, visible to the naked eye and a beautiful sight through binoculars or a small telescope.

Latin name:	Perseus
Latin genitive:	Persei
Abbreviation:	Per
Meaning:	Perseus
Visibility:	Northern autumn, southern spring

Cassiopeia

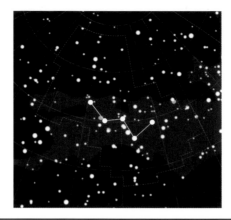

One of the most familiar constellations in the sky, Cassiopeia is a compact W-shaped group of stars in the northern Milky Way. It is circumpolar for much of the northern hemisphere, and lies on the opposite side of the pole star from the Great Bear Ursa Major. The constellation represents Queen Cassiopeia, the vain mother of Andromeda, seated on her throne. Gamma (γ) Cassiopeiae, at the centre of the 'W', is an unpredictable variable – a rapidly spinning blue-white star with ten times the mass of the Sun, surrounded by a shell of material that it has flung out from its own equator. At times, it can outshine Schedar (α Cas) and Caph (β Cas). The constellation's other highlights are its open star clusters, the brightest of which is M52. Easily picked up with binoculars or a small telescope, it contains about 200 stars in a region some 20 light years across. From Earth, it is about half as wide as a full Moon.

Latin name:	Cassiopeia
Latin genitive:	Cassiopeiae
Abbreviation:	Cas
Meaning:	Queen Cassiopeia
Visibility:	Northern hemisphere circumpolar

Orion

The large and distinctive constellation of Orion represents a great hunter from Greek myth, the doomed lover of Diana, goddess of the hunt. He is depicted facing the charging bull Taurus, with raised shield and club, and his dogs, Canis Major and Minor, at his back. Orion's shoulder and foot are marked by two of the brightest stars in the sky – the red giant Betelgeuse* and the brilliant blue-white Rigel, or Beta (β) Orionis. Rigel is actually the brighter of the two stars at magnitude 0.1 – it is a brilliant supergiant 770 light years away and shining with the luminosity of perhaps 66,000 Suns. Orion's belt is marked by three second-magnitude stars in a row, and below these hangs the hunter's sword, incorporating the multiple star Sigma Orions (single to the naked eye, double through binoculars, and quadruple through a small telescope) and the beautiful star-forming nebula M42*.

Latin name:	Orion
Latin genitive:	Orionis
Abbreviation:	Ori
Meaning:	Orion
Visibility:	Northern winter, southern summer

Cepheus

This indistinct constellation represents King Cepheus of Ethiopia – husband of Cassiopeia and father of Andromeda in the cycle of myths connected with Perseus. Cepheus's chief claim to fame is that it contains several famous variable stars, of which the best known is the pulsating yellow supergiant Delta Cephei*. Delta's variations are easily predictable and can be followed with the naked eye. By contrast, Beta (β) Cephei or Alfirk is a hot blue star that varies by 0.1 magnitude, and has a complex series of overlapping cycles, with periods around 4.5 hours. Mu (μ) Cephei, meanwhile, is an even less predictable red supergiant, one of the most extreme stars in the sky with the energy output of 350,000 Suns and a diameter larger than the orbit of Jupiter. It lies 5000 light years away, shines with an average magnitude of 4.0, and is destined to die in a supernova explosion in the relatively near future.

Latin name:	Cepheus
Latin genitive:	Cephei
Abbreviation:	Cep
Meaning: King	Cepheus
Visibility:	Northern hemisphere circumpolar

Lynx

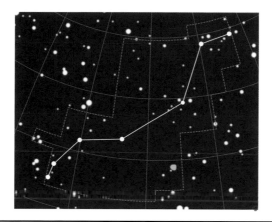

This chain of faint stars to the southwest of Ursa Major was added to the sky by Polish astronomer Johannes Hevelius in the 1680s. It bears no resemblance to a cat, and in this case it was not intended to: the name is an astronomical pun, since Hevelius commented that only the 'lynx-eyed' would be able to spot its stars. The most interesting of these faint stars is undoubtedly 12 Lyncis, an attractive multiple star that can be split by a small telescope into a primary of magnitude 4.9 with a wide companion of magnitude 7.1. A larger instrument will split the primary into a tight double of magnitudes 5.4 and 6.0. Lynx's other highlight is an intriguing globular cluster, faint at magnitude 10.4 and best tracked down with a small telescope. The 'Intergalactic Wanderer' lies 295,000 light years away, far above the Milky Way, and is probably an escapee from another galaxy rather than a satellite of our own.

Latin name:	Lynx
Latin genitive:	Lyncis
Abbreviation:	Lyn
Meaning:	Lynx
Visibility:	Northern spring, southern autumn

Libra

The only zodiac constellation to represent an inanimate object, Libra was once not even a constellation in its own right – it was simply Chelae Scorpionis, the extended claws of neighbouring Scorpius. Today, however, it is seen as a set of scales, held aloft by Virgo in her guise of Astreia, goddess of justice. Alpha (α) Librae is Zubenelgenubi, or 'southern claw' (a hangover from the constellation's past). Binoculars will reveal that this is a wide double star, its two stars separated by a light month or more, and orbiting each other in perhaps 200,000 years. The brighter member of the system is itself a spectroscopic binary. Beta (β) Lib, or Zubenelschamali, is a single star, slightly brighter than Alpha at magnitude 2.6, and can appear greenish to some observers. Iota (ι) Lib is a complex quintuple star that reveals more elements when viewed through more powerful instruments. NGC 5897 is a fairly faint and distant globular cluster.

Latin name:	Libra
Latin genitive:	Librae
Abbreviation:	Lib
Meaning:	Scales
Visibility:	Northern summer, southern winter

Gemini

The constellation of the Twins gets its name from its two neighbouring bright stars, Castor and Pollux. In Greek myth, these two were known as the Dioscuri – members of Jason's crew of Argonauts. In reality, the stars are unrelated, though both are relatively close to Earth, at 52 and 34 light years respectively. However, Alpha (α) Geminorum or Castor is a famous multiple in its own right. A small telescope will split it into a pair of blue-white stars of magnitudes 1.9 and 2.9, orbited by a nearby red dwarf of magnitude 9.3. Each of these stars is itself binary, making Castor a six-star system. Pollux, in contrast, is a lone orange giant, though it is orbited by at least one planet, with a mass of perhaps three Jupiters. Elsewhere in the constellation, M35 is an attractive open star cluster, just visible with the naked eye, and NGC 2392 is the Eskimo Nebula, a planetary nebula for small telescopes.

Latin name:	Gemini
Latin genitive:	Geminorum
Abbreviation:	Gem
Meaning:	Twins
Visibility:	Northern winter, southern summer

Cancer

This faint triangle of stars is far from eyecatching, but can easily be located since it lies between the bright stars Castor and Pollux (in Gemini) and Regulus (in Leo). It represents a crab, crushed underfoot by the hero Hercules during his struggle with the dragon Draco. Its most attractive feature is the star cluster M44, known as The Beehive or Praesepe*. This is easily visible to the naked eye in dark skies, and a beautiful sight for binoculars, which reveal its rich fields of fairly bright stars. Another cluster, M67, lies much further away (2700 light years from Earth) and is visible only through binoculars. However, it is worth tracking down because it is very unusual – most open clusters disintegrate quite quickly, scattering their stars across space in a few tens of millions of years. M67, in contrast, has held together for about four billion years.

Latin name:	Cancer
Latin genitive:	Cancri
Abbreviation:	Cnc
Meaning:	Crab
Visibility:	Northern spring, southern autumn

Vela

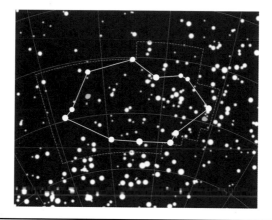

This misshapen oval of stars represents the sail of Jason's great ship, the Argo. With a rich seam of the southern Milky Way running across it, the constellation is rich in interesting stars and deep sky objects. Regor or Gamma (γ) Velorum, the brightest star, is a complex multiple containing six stars in all. Four elements can be seen with a small telescope, but the most fascinating is the bright primary star at magnitude 1.8. Analysis of its spectrum has revealed a monstrous binary – a hot blue star with the mass of 30 Suns, with a 'Wolf-Rayet' companion that now weighs the same as 10 Suns. Delta (δ) Vel is another complex multiple, while IC 2391 is a bright naked-eye open cluster, and NGC 3201 an attractive globular for binoculars. Two expanding supernova remnants also overlay much of the constellation – the 'Vela Supernova Remnant' is a mere 10,000 years old, while the fainter Gum Nebula formed a million years ago.

Latin name:	Vela
Latin genitive:	Velorum
Abbreviation:	Vel
Meaning:	Sail
Visibility:	Northern winter, southern summer

Scorpius

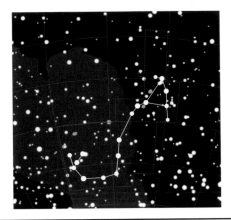

This bright zodiac constellation commemorates the scorpion that killed the great hunter Orion – it lies on the opposite side of the sky for that reason, crossing a rich band of the Milky Way. Its brightest star is Antares*, whose name means 'the rival to Mars' – a red supergiant 600 light years away. Antares is flanked by Sigma and Tau Scorpii, making it fairly unmistakeable. The constellation contains several bright star clusters, of which the most impressive are the globular M4 and the open cluster M7, both visible to the naked eye in a clear dark sky. It is also rich in multiple stars. Beta (β) Scorpii, or Acrab, consists of at least five stars in orbit around each other, and small telescopes will reveal the two brilliant blue primaries of magnitudes 2.6 and 4.9. Nu (ν) Scorpii appears double through small telescopes, while larger ones show that each of its single stars is in fact double again.

Latin name:	Scorpius
Latin genitive:	Scorpii
Abbreviation:	Sco
Meaning:	Scorpion
Visibility:	Northern summer, southern winter

Carina

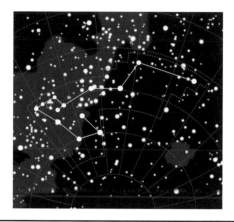

The constellation of the Keel is one-third of the huge and ancient Argo Navis, broken up by Nicolas de Lacaille in the 1750s. It is unmistakeable thanks to the presence of Canopus, the second brightest star in the entire sky, and contains some beautiful starfields and nebulosities – notably the Carina Nebula* and the Homunculus Nebula embedded within it, surrounding the dying monster star Eta (η) Carinae*. Canopus is a yellow-white supergiant 315 light years away. Since it shines at magnitude -0.7 and emits its light at roughly the same wavelengths as the Sun, it must be about 15,000 times as luminous as our star – or 600 times as luminous as Sirius. Elsewhere, Carina is rich in star clusters: the Southern Pleiades (IC 2602) are the most impressive, a group of about 60 stars some 480 light years away, of which the brightest is Theta (ϑ) Carinae at magnitude 2.7.

Latin name:	Carina
Latin genitive:	Carinae
Abbreviation:	Car
Meaning:	Keel
Visibility:	Northern winter, southern summer

Monoceros

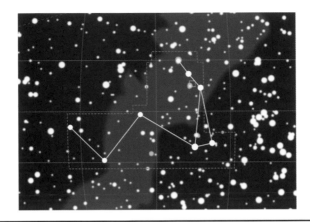

The constellation of the Unicorn is a large, faint W-shape that happens to overlie a rich area of the Milky Way. It may have been invented to fill in a gap in the sky by Dutch theologian and astronomer Petrus Plancius around 1613, but it might also be older, with perhaps a medieval Arabic origin. Beta (β) Monocerotis is a beautiful triplet of blue stars, each about 6–7 times the mass of the Sun, easily separated with a small telescope. M50 is a colourful star cluster for binoculars, with bright members that range from hot blue stars to cool red giants. NGC 2264 is a naked-eye cluster containing the monstrous S Monocerotis, a binary consisting of twin stars with 20–30 solar masses each, appearing to the naked eye as a single star of magnitude 4.7. NGC 2244, meanwhile, is another naked-eye cluster at the heart of the extensive, though faint Rosette Nebula.

Latin name:	Monoceros
Latin genitive:	Monocerotis
Abbreviation:	Mon
Meaning:	Unicorn
Visibility:	Northern winter, southern summer

Sculptor

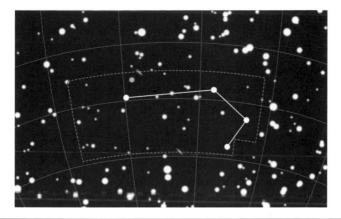

One of many constellations invented by French astronomer Nicolas de Lacaille, Sculptor supposedly represents, in a few unimpressive stars, an entire sculptor's workshop. Although the constellation itself is less than convincing, its chief attractions are two bright, nearby galaxies – members of the Sculptor Group, the nearest family of galaxies to our own Local Group. NGC 253, the Silver Coin Galaxy*, is an almost edge-on spiral visible through binoculars, while NGC 55* is a curious hybrid galaxy just beyond the Local Group at a distance of seven million light years, and easily observed through a small telescope. Closer to home, R Sculptoris is a 'semi-regular' variable star – a swollen red giant that appears all the redder because it is surrounded by a shell of carbon that absorbs short wavelengths and bluer colours. It pulsates between magnitudes 5.8 and 7.7 in a little over a year, and is best tracked with binoculars.

Latin name:	Sculptor
Latin genitive:	Sculptoris
Abbreviation:	Scl
Meaning:	Sculptor
Visibility:	Northern autumn, southern spring

Phoenix

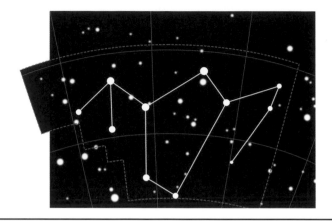

A 'southern bird' added to the sky by Dutch traders and navigators Pieter Dirkszoon Keyser and Frederick de Houtman during their travels in the East Indies in the 1590s, Phoenix represents the mythical firebird that regenerates from its own ashes. Intriguingly, earlier Chinese astronomers appear to have seen a similar creature, their own symbol of death and regeneration, in the same group of stars, raising the possibility that Keyser and de Houtman were influenced by local tales. Alpha (α) Phoenicis, or Ankaa, is an orange giant with about 2.5 times the mass of the Sun, and an unseen companion in a 10-year orbit that affects its spectrum. Zeta (ζ) Phoenicis is a tightly bound multiple star, with a companion (divisible in large telescopes) of magnitude 6.9, orbiting an inseparable pair of blue-white stars that eclipse each other as they orbit every 40 hours, causing the brightness of the system to briefly dip from its usual magnitude 4.4.

Latin name:	Phoenix
Latin genitive:	Phoenicis
Abbreviation:	Phe
Meaning:	Phoenix
Visibility:	Northern autumn, Southern spring

Canes Venatici

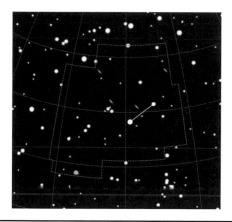

Tucked under the tail of the Great Bear Ursa Major, Canes Venatici is an easily overlooked constellation marked by just one bright star. It represents the hunting dogs of the herdsman Boötes, perpetually chasing the bears around the north celestial pole, and was first catalogued in the 1680s by Polish astronomer Johannes Hevelius. Alpha (α) CVn is Cor Caroli, an attractive double star of magnitudes 2.9 and 5.5, easily split with a small telescope. Named in honour of the executed British King Charles I, it is the prototype for a class of variable stars that change their brightness as they rotate, suggesting their surfaces are 'blotchy' and covered in enormous sunspots. Elsewhere in the constellation lies the fine globular cluster M3, a ball of half a million stars some 34,000 light years away, easily spotted with binoculars or a small telescope. Another highlight is the Whirlpool Galaxy M51*, best observed through binoculars or larger telescopes.

Latin name:	Canes Venatici
Latin genitive:	Canun Venaticorum
Abbreviation:	CVn
Meaning:	Hunting Dogs
Visibility:	Northern spring, Southern autumn

Aries

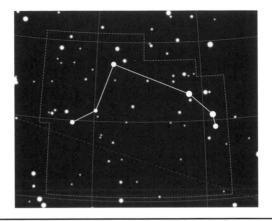

This zodiac constellation, representing the ram that bore the golden fleece sought by Jason and the Argonauts, has a somewhat confusing pattern, and is best found by looking midway between the more obvious groupings of Andromeda and Taurus. In ancient times, the Sun crossed the celestial equator here, heralding the start of summer in the northern hemisphere. This crossing point, the basis for our celestial coordinate system, is still known as the first point of Aries, even though precession has now carried it into Pisces. Alpha (α) Arietis is Hamal, a typical orange giant star, 66 light years away and shining at magnitude 2.0. And Gamma (γ) Ari, or Mesarthim, is an attractive multiple – a small telescope will transform the single magnitude 4.1 'star' into an evenly matched white pair of magnitudes 4.6 and 4.7, orbited by a more distant orange star of magnitude 9.6. The entire system lies 205 light years away.

Latin name:	Aries
Latin genitive:	Arietis
Abbreviation:	Ari
Meaning:	Ram
Visibility:	Northern autumn, southern spring

Capricornus

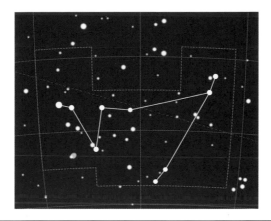

This constellation supposedly represents a bizarre hybrid of goat and fish – the creature into which the god Pan transformed in his efforts to escape the sea monster Typhon. Its pattern is obscure, however, and it is best located relative to the brighter stars of Sagittarius and Piscis Austrinus. Its most intriguing features are probably its several bright multiple star systems. Alpha (α) Capricorni, or Algiedi, is a line-of-sight double divisible with the naked eye, consisting of a yellow giant 108 light years away, and a yellow supergiant 635 light years away. A small telescope will show that each star is itself a true binary with a faint companion. Some 330 light years away is Beta (β) Capricorni, an extremely complex multiple, consisting of at least seven stars, of which the brightest two can be separated with binoculars. M30, meanwhile, is an attractive globular cluster with a very dense core and loose outer layers of stars, 26,000 light years distant.

Latin name:	Capricornus
Latin genitive:	Capricorni
Abbreviation:	Cap
Meaning:	Sea Goat
Visibility:	Northern summer, southern winter

Fornax

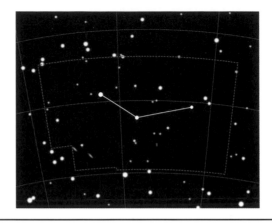

Fornax is a typically faint constellation invented by French astronomer Nicolas de Lacaille during his observations from the Cape of Good Hope in the 1750s. This was originally Fornax Chemica, the Chemical Furnace, since Lacaille named many of his new star patterns after the scientific instruments of the Enlightenment – the intellectual movement of the eighteenth century. Alpha (α) Fornacis is a binary star 42 light years away, easily split with binoculars or a small telescope to reveal a pair of yellow stars that neatly bracket our Sun in terms of mass and brightness. The magnitude 4.0 primary has 25 per cent more mass than the Sun and is four times more luminous, while its companion, of magnitude 6.5, has just 75 per cent of the Sun's mass, and half its luminosity. NGC 1316* is a large lenticular galaxy, visible through small telescopes at magnitude 9.4. It is a strong source of radio waves also known as Fornax A, and an outlying member of the huge Fornax galaxy cluster.

Latin name:	Fornax
Latin genitive:	Fornacis
Abbreviation:	For
Meaning:	Furnace
Visibility:	Northern winter, southern summer

Coma Berenices

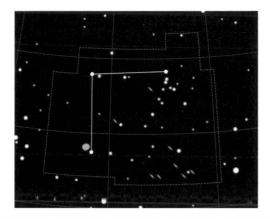

Although this constellation's pattern is indistinct, it is easy to locate because it lies directly to the east of Leo. Under a dark sky, its numerous faint stars merge into V-shaped strands that led early astronomers to see it as the tuft on the lion's tail. Late in antiquity, it was renamed to honour Queen Berenice of Egypt, who sacrificed her golden hair to the gods in thanks for her husband's safe return from a military campaign. The Coma Star Cluster, also known as Melotte 111, lies 290 light years from Earth and is one of the closest open clusters to Earth. It makes a beautiful target for binoculars. Coma is also scattered with galaxies – the brightest of which are members of the Virgo Cluster (the nearest major galaxy cluster), spilling over from the neighbouring constellation to the south. A more distant group, the Coma Galaxy Cluster, lies about 320 million light years away, close to Beta (β) Comae.

Latin name:	Coma Berenices
Latin genitive:	Comae Berenices
Abbreviation:	Com
Meaning:	Queen Berenice's Hair
Visibility:	Northern spring, southern autumn

Canis Major

Orion's larger hunting dog is one of the sky's brightest constellations. It not only contains the 'Dog Star' Sirius*, the brightest star of all at magnitude -1.46, but also several other bright stars and a rich swathe of the Milky Way. While Sirius is just an average star that happens to be nearby, Beta (β) Canis Majoris or Mirzam is a blue giant shining at magnitude 2.0 despite a distance of 500 light years. Delta (δ) CMa or Wezen is an even more brilliant yellow supergiant – 1800 light years away, yet still slightly brighter than Mirzam. M41 and NGC 2362 are contrasting open clusters, both visible to the naked eye and ideal targets for binoculars or a small telescope. M41 is almost 200 million years old, and its bright stars are red and orange giants, while NGC 2362, around Tau (τ) CMa, is just 25 million years old, and still dominated by brilliant blue and white stars.

Latin name:	Canis Major
Latin genitive:	Canis Majoris
Abbreviation:	CMa
Meaning:	Great Dog
Visibility:	Northern winter, southern summer

Pavo

Pavo, the Peacock, is one of the 'southern birds', a group of constellations that includes Grus, the Crane and Phoenix, and which was added to the sky by the Dutch navigators Pieter Dirkszoon Keyser and Frederick de Houtman in the late 1500s. Although its pattern is indistinct, the constellation is easy to spot because of its single bright star, Alpha (α) Pavonis, sometimes known as 'Peacock'. This hot blue star on the 'main sequence' of stellar evolution lies 183 light years away. About five times the mass of the Sun, it produces 2000 times its energy. Kappa Pavonis, though fainter, is one of the brightest Cepheid variables (yellow pulsating supergiants similar to Delta Cephei). It varies between magnitudes 3.9 and 4.8 in a 9.1-day cycle. NGC 6752 is one of the closest globular clusters to Earth, 13,000 light years away and visible to the naked eye as a moon-sized smudge of light.

Latin name:	Pavo
Latin genitive:	Pavonis
Abbreviation:	Pav
Meaning:	Peacock
Visibility:	Southern hemisphere circumpolar

Grus

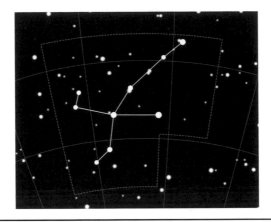

The constellation of the Crane is one of several 'southern birds' added to the sky by Dutch navigators Pieter Dirkszoon Keyser and Frederick de Houtman in the late 1500s. It lies just to the south of Piscis Austrinus with its bright star Fomalhaut, and its stars were once seen as part of that more ancient constellation. Nevertheless, its skewed cross-shape does bear some resemblance to a flying bird. The brightest star, magnitude 1.7 Alpha (α) Gruis, is also called Alnair, which means simply 'bright one' but comes from a longer Arabic phrase meaning 'the bright one in the fish's tail'. It is a hot blue-white star 101 light years away, with the mass of four Suns and a surface temperature of about 13,000°C (23,000°F). Beta (β) Gruis is a red giant 170 light years away, and an unpredictable variable star that fluctuates between magnitudes 2.0 and 2.3.

Latin name:	Grus
Latin genitive:	Gruis
Abbreviation:	Gru
Meaning:	Crane
Visibility:	Northern autumn, southern spring

Lupus

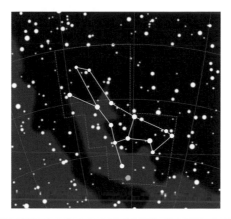

Although its Latin name means 'wolf', Lupus for a long time represented an indistinct animal impaled on the spear of the neighbouring Centaurus. Of its several star clusters, the finest globular is NGC 5986, and the best open cluster NGC 5822, but both require binoculars in order to spot them. Alpha (α) Lupi or Kakkab is a hot blue star with a surface temperature of 21,0000°C (38,0000°F). It is one of many related stars of similar age, scattered across this area of the sky in the 'Scorpius-Centaurus OB association'. The stars were born in an open cluster about 20 million years ago, and appear well scattered only because they are relatively nearby – about 500 light years from Earth. Mu (μ) Lupi (magnitude 4.3) reveals a magnitude 7.2 companion through small telescopes, while larger instruments show that the brighter star is itself a close binary of blue-white stars of magnitudes 5.1 and 5.2.

Latin name:	Lupus
Latin genitive:	Lupi
Abbreviation:	Lup
Meaning:	Wolf
Visibility:	Northern spring, southern autumn

Sextans

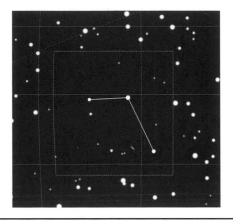

This constellation is one of a handful invented by the Polish astronomer Johannes Hevelius, who published them in his posthumous star catalogue, *Uranometria,* in 1687. Its broad triangle of stars represents a sextant, a navigational instrument used by sailors for measuring the precise position of the Sun in order to find their own location on Earth. The constellation has few objects of interest for amateur astronomers. Its brightest stars – Alpha (α) at magnitude 4.5 and Beta (β) and Gamma (γ) at magnitude 5.1 – are distant white or blue-white main-sequence stars, all several hundred light years away. Perhaps Sextans' most rewarding object is NGC 3115, one of the closest large elliptical galaxies to Earth. This elongated ball of ageing red and yellow stars lies 14 million light years from Earth, and shines at magnitude 8.5 – bright enough to spot through binoculars, although a telescope is needed to reveal any structure.

Latin name:	Sextans
Latin genitive:	Sextantis
Abbreviation:	Sex
Meaning:	Sextant
Visibility:	Northern spring, southern autumn

Tucana

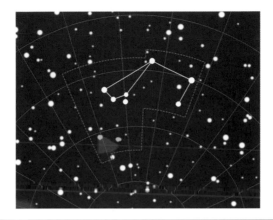

One of several 'southern birds' invented by Dutch navigators Pieter Dirkszoon Keyser and Frederick de Houtman during their voyages in the late sixteenth century, Tucana represents a Toucan. Its stars are relatively faint but it is still easy to locate, since it lies just to the west of brilliant Achernar in Eridanus, and contains the Small Magellanic Cloud* (SMC) within its boundaries. This satellite galaxy of our own Milky Way, more distant and slightly fainter than the Large Magellanic Cloud of Dorado is nevertheless visible to the naked eye, and a rewarding sight through binoculars or a small telescope. Elsewhere in the constellation lies 47 Tucanae*, one of the sky's best globular clusters, while another globular, NGC 362, lies on the edge of the SMC as seen from Earth, though it is actually a foreground object. Beta Tucanae, meanwhile, is a complex multiple star – a combination of line-of-sight effects and physical multiples.

Latin name:	Tucana
Latin genitive:	Tucanae
Abbreviation:	Tuc
Meaning:	Toucan
Visibility:	Southern hemisphere circumpolar

Indus

The constellation of Indus is usually taken to represent a North American 'Indian', despite the fact that its inventors, Dutch navigators Keyser and de Houtman, first devised it while travelling in the 'East Indies' of Southeast Asia. As with Phoenix*, there is some evidence that they might have been influenced by an earlier association of these stars with an exotic foreigner, since the Chinese knew the same group of stars as 'The Persian'. Alpha (α) Indi still bears this Chinese name – it is a triple star system dominated by an orange giant of magnitude 3.1, with two dwarf companions visible through larger telescopes. However, Epsilon (ε) Indi is the constellation's most fascinating star – just 11.8 light years away and Sunlike (with a mass of 0.8 Suns and a luminosity 20 per cent of our star's), it is seen as a potential nearby home for extraterrestrial life.

Latin name:	Indus
Latin genitive:	Indi
Abbreviation:	Ind
Meaning:	Indian
Visibility:	Northern summer, southern winter

Octans

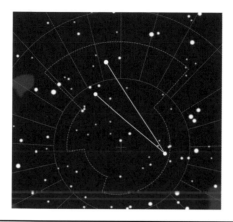

The constellation that incorporates the south celestial pole is certainly no rival for Ursa Minor in northern skies. Octans is another invention of French astronomer Nicolas Louis de Lacaille, who filled the gaps in the southern sky while working at Cape Town during the 1750s. It represents an octant – a navigation and surveying instrument that was precursor to the sextant, and is marked out in a triangle of unimpressive fourth-magnitude stars. Lambda Octantis is probably the constellation's most interesting object – a double for small telescopes consisting of yellow and white stars of magnitudes 5.4 and 7.7, and 435 light years away. The southern pole star itself is Sigma (σ) Octantis – an uninspiring yellow-white star 270 light years from Earth, and shining at magnitude 5.4. With a mass twice that of the Sun, Sigma has recently left the main sequence and has begun to swell into a giant.

Latin name:	Octans
Latin genitive:	Octantis
Abbreviation:	Oct
Meaning:	Octant
Visibility:	Southern hemisphere circumpolar

Lepus

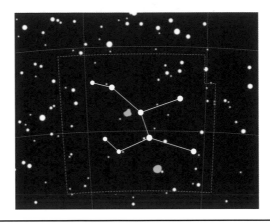

This distinctive bow-tie shaped pattern of stars directly to the south of Orion represents a hare fleeing the approach of the hunter's dogs Canis Major and Minor. Although relatively small and often overlooked in favour of its brighter neighbours, it contains several interesting stars and other objects. Alpha (α) Leporis or Arneb is a white supergiant star, 1300 light years away but still quite prominent at magnitude 2.6. R Leporis* (also known as Hind's Crimson Star), is a beautiful red giant variable similar to Mira in Cetus. It oscillates between magnitudes 5.5 and 11.7 in a 430-day period, so its changes are best tracked with a small telescope. M79 is a fairly faint globular cluster at magnitude 7.9, located in a curious part of the sky compared to other globulars, which suggests it may be a refugee from a dwarf galaxy captured and absorbed by the Milky Way.

Latin name:	Lepus
Latin genitive:	Leporis
Abbreviation:	Lep
Meaning:	Hare
Visibility:	Northern winter, southern summer

Lyra

This compact constellation is made obvious by the presence of Vega*, one of the brightest stars in the sky. The pattern represents a lyre – an ancient musical instrument played by various mythological bards, including Orpheus during his journey through the underworld, and Arion, the bard rescued by the dolphin Delphinus at the command of the god Poseidon. Despite its small size, Lyra includes many interesting objects. Sheliak*, or Beta (β) Lyrae, is an unusual eclipsing binary star, while Epsilon (ε) Lyrae* is an impressive multiple star. RR Lyrae*, meanwhile, is the prototype for a group of variable stars also called 'dwarf Cepheids'. It has an average magnitude of 7.5 and is easily found with binoculars, as is the beautiful Ring Nebula M57*, one of the brightest planetary nebulae in the sky.

Latin name:	Lyra
Latin genitive:	Lyrae
Abbreviation:	Lyr
Meaning:	Lyre
Visibility:	Northern summer, southern winter

Crater

This fairly distinctive constellation – a 'bow tie' of stars to the south of Leo – lies on the back of Hydra, the water snake, right alongside Corvus, the crow. An ancient legend links all three: one day, the god Apollo sent his messenger, the crow, to fetch water in his cup. Along the way, the bird got distracted by a ripe fig tree, and when he returned late, he carried a water snake in his claws, claiming that this creature had been blocking the well. Apollo saw through the crow's lie, and in his anger threw bird, cup and snake into the heavens, where they can still be seen. Unfortunately, the constellation is a fairly barren one, with little to interest amateur astronomers. The brightest star is Delta (δ) Crateris at magnitude 3.6 – another of the sky's plentiful naked-eye orange giants. Magnitude 4.1 Gamma (γ) Crt has a binary companion of magnitude 9.6, visible with a small telescope.

Latin name:	Crater
Latin genitive:	Crateris
Abbreviation:	Crt
Meaning:	Cup
Visibility:	Northern spring, southern autumn

Columba

Lying just to the southwest of Canis Major, this constellation was invented by Dutch astronomer and theologian Petrus Plancius in the late sixteenth century. In keeping with his programme to Christianize the skies, Plancius intended it to represent the dove sent out by Noah from his ark. In this reading of the sky, Jason's ship, the *Argo*, was transformed into the Ark itself, but the dove can equally be seen as the bird that flew ahead of the *Argo* to guide it into the Black Sea. Alpha (α) Columbae is Phact, a blue-white rapidly spinning star that is beginning its evolution toward giant status. Like Gamma Cassiopeiae, it is slightly variable and surrounded by a shell of gases flung out from its own equator. Binoculars or a small telescope will reveal the globular cluster NGC 1851, some 39,000 light years away but still luminous enough to shine at magnitude 7.1.

Latin name:	Columba
Latin genitive:	Columbae
Abbreviation:	Col
Meaning:	Dove
Visibility:	Northern winter, southern summer

Vulpecula

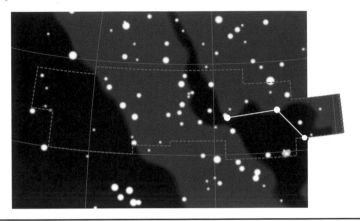

This small constellation was added to the sky by Johannes Hevelius around 1687. It was originally known as Vulpecula et Anser, the Fox and Goose, but has since been relegated to simply a fox, although Anser is still commemorated in the name of Alpha (α) Vulpeculae. Despite its size, Vulpecula has several interesting objects. M27 is the Dumbbell Nebula, one of the largest and brightest planetary nebulae in the skies. It can be spotted with binoculars on dark nights, and is a good target for small telescopes, with a distinctive bright bar created as the central star flings off material around its equator, which happens to be edge-on to our line of sight. Brocchi's Cluster, also known as the Coathanger, appears to be a bright naked-eye cluster that reveals about 40 stars through binoculars. However, it is actually a line-of-sight cluster caused by the chance overlap of several small groups of stars at varying distances.

Latin name:	Vulpecula
Latin genitive:	Vulpeculae
Abbreviation:	Vul
Meaning:	Fox
Visibility:	Northern summer, southern winter

Ursa Minor

The Lesser Bear has a shape that roughly mimics the Plough or Big Dipper pattern in Ursa Major, but its chief claim to fame is the way that it pivots around the sky, suspended by the bright star at the end of its tail, Polaris. The renowned pole star of northern skies is a yellow supergiant of magnitude 2.0, which is 2400 light years away and just happens to line up more or less exactly with the point in space at which Earth's north pole is currently pointing. It will not always be the pole star, however, since our planet's axis slowly wobbles like a spinning top, pointing in different directions over a cycle of 25,800 years. Interesting in its own right, Polaris has been closely studied for centuries, and so we know that while it used to be a variable star similar to Delta Cephei*, it may also have been brighter than it is today.

Latin name:	Ursa Minor
Latin genitive:	Ursae Minoris
Abbreviation:	UMi
Meaning:	Little Bear
Visibility:	Northern hemisphere circumpolar

Telescopium

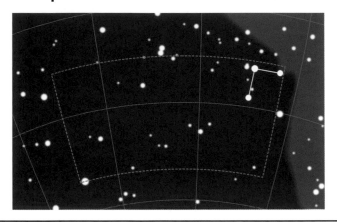

This constellation, invented by French observer Nicolas de Lacaille in the 1750s, is a disappointing tribute to the astronomer's most important instrument – a barren rectangle of sky with the supposed pattern of stars jammed awkwardly into its northwest corner. It was not always this way, though: in order to build his constellation, Lacaille 'borrowed' stars from its neighbours, including Coronae Australis, Sagittarius and Scorpius. This was not such a problem in an age when the constellations were defined purely as patterns made by the stars, but when astronomers decided in 1929 to redefine them as areas of sky, so that every object in the heavens would fall within one or another, Telescopium's stolen stars were returned to their rightful owners, leaving the constellation in its current pitiful state. Its most notable star is Delta (δ) Telescopii, a double star that is a chance alignment of stars at quite different distances from Earth.

Latin name:	Telescopium
Latin genitive:	Telescopii
Abbreviation:	Tel
Meaning:	Telescope
Visibility:	Northern summer, southern winter

Horologium

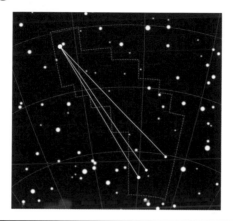

A typically faint and obscure invention of French astronomer Nicolas de Lacaille, Horologium represents a pendulum clock, though in reality it is little more than a long triangle of unimpressive stars alongside the southern reaches of the long river constellation Eridanus. Alpha (α) Horologii is an orange giant 117 light years away and shining at magnitude 3.9. More interesting, though harder to spot, is the variable star R Horologii. A 'long-period variable' similar to the famous Mira* in Cetus , it has far more extreme variations that take it from magnitude 14 to magnitude 4 and back again in a cycle of 407 days – a variation in actual luminosity of 400,000 per cent. Further afield, NGC 1512 is a magnificent barred spiral galaxy, 30 million light years away and just visible through a small telescope at magnitude 11.1, though better seen in larger instruments. It is about 70,000 light years across, and has a small elliptical galaxy companion, NGC 1510.

Latin name:	Horologium
Latin genitive:	Horologii
Abbreviation:	Hor
Meaning:	Clock
Visibility:	Northern winter, southern summer

Pictor

This obscure star pattern is supposed to represent a painter's easel, and like many southern constellations, it was a late addition to the sky, invented by French astronomer Nicolas Louis de Lacaille in the 1750s. Its brightest star is the white Alpha (α) Pictoris, magnitude 3.3 and 99 light years from Earth, but magnitude 3.9 Beta (β) Pictoris* is far better known, since this was the first star found to be surrounded by a potentially planet-forming disk of gas and dust. Delta (δ) Pictoris, meanwhile, is an eclipsing binary – a normal-looking blue-white star of magnitude 4.7, which dips briefly to magnitude 4.9 every 40 hours. Delta is very remote at 1650 light years from Earth, so its components must be extremely luminous. By analyzing the light curve obtained by plotting the system's changes over time, astronomers have concluded that one star has twice the mass of the other.

Latin name:	Pictor
Latin genitive:	Pictoris
Abbreviation:	Pic
Meaning:	Easel
Visibility:	Northern winter, southern summer

Piscis Austrinus

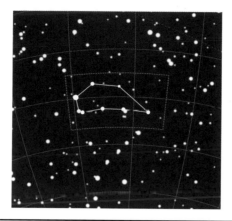

The 'southern fish' is a fairly obscure constellation dominated by its brightest star, the nearby white main-sequence Fomalhaut*. It is one of the 48 original constellations of the ancient Greeks, despite lying well into the sky's southern hemisphere. It was said to represent the parent of the two fish in Pisces, and was shown swimming up the stream of water poured by the nearby Aquarius. The next brightest star after Fomalhaut is magnitude 4.3 Beta (b) Piscis Austrini, a white star 148 light years away, with a companion of magnitude 7.7 visible through a small telescope. Gamma (g) PsA is also double, more distant at 220 light years, and more challenging since the magnitude 8.0 secondary star lies closer to the magnitude 4.3 primary and so tends to be lost in its glare. Nevertheless, medium-sized telescopes should have enough power to split the pair.

Latin name:	Piscis Austrinus
Latin genitive:	Piscis Austrini
Abbreviation:	PsA
Meaning:	Southern Fish
Visibility:	Northern autumn, southern spring

171

Hydrus

This compact zigzag of stars leading towards the south celestial pole represents a second water snake in the sky – far smaller than the extensive Hydra that swims just below the celestial equator. It was invented by Dutch navigators Keyser and de Houtman in the late 1500s. Beta (β) Hydri is the brightest star at magnitude 2.8. A sunlike star just 24 light years away, it is slightly heavier than our own Sun, and a little older at 6.7 billion years. Its extra mass means that it has aged more quickly than the Sun will – it has already exhausted the supply of hydrogen in its core, and has begun to swell into a giant as hydrogen fusion spreads out from its core into the surrounding material. Alpha (α) Hyi, meanwhile, is a more massive and luminous white main sequence star – further away at 71 light years, but only slightly fainter than Beta, with a magnitude of 2.9.

Latin name:	Hydrus
Latin genitive:	Hydri
Abbreviation:	Hyi
Meaning: Little	Water Snake
Visibility:	Southern hemisphere circumpolar

Antlia

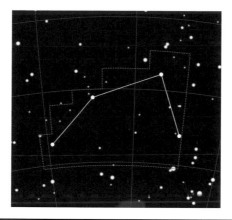

Situated in a fairly barren area of sky between Hydra and the brighter stars of Vela, this faint and formless constellation was added to the sky by French astronomer Nicolas Louis de Lacaille in the 1650s. He intended it to represent an air pump, a scientific device developed by the British scientist Robert Hooke and French inventor Denis Papin in the late seventeenth century. Its brightest star, Alpha (a) Antliae, is a paltry magnitude 4.3 – an orange giant about 365 light years from Earth. Zeta (ζ) Antliae, meanwhile, is an interesting double star, revealed with binoculars to contain twin white components of magnitues 5.8 and 5.9. These two stars lie within a couple of light years of each other, but their other properties suggest they are independent systems, and their alignment is a mere chance encounter. A small telescope will show that the western star of the pair is itself a true binary, consisting of stars with magnitudes 6.2 and 7.0.

Latin name:	Antlia
Latin genitive:	Antliae
Abbreviation:	Ant
Meaning:	Air Pump
Visibility:	Northern spring, southern autumn

Ara

This small grouping, embedded in the Milky Way to the south of Scorpius, is one of the most obscure constellations to originate in ancient times. Classical astronomers saw it as an altar on which the Greek gods swore their oaths. Magnitude 5.2 Mu (μ) Arae is a fairly sunlike yellow star 50 light years away, home to a complex system of at least four planets, ranging from ten times the mass of Earth to twice the mass of Jupiter. NGC 6193, meanwhile, is an open cluster 4300 light years away, on the limit of naked-eye visibility, dominated by a blue-white giant of magnitude 5.7. Its stars are only a million years old, and the cluster is still embedded in the emission nebula from which it formed, NGC 6188. NGC 6397, in contrast, is an ancient globular cluster, quite nearby for one of its type, at 7200 light years. Easily found with binoculars, it has a highly concentrated core.

Latin name:	Ara
Latin genitive:	Arae
Abbreviation:	Ara
Meaning:	Altar
Visibility:	Northern summer, southern winter

Leo Minor

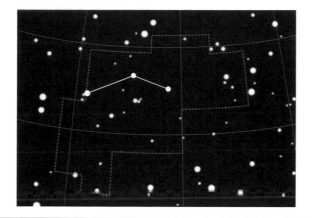

Unlike its larger cousin, Leo Minor bears no resemblance to a lion. In fact, it has no distinctive pattern at all. The constellation was invented by Polish astronomer Johannes Hevelius in the 1680s, apparently to fill a convenient gap in the sky. It has no stars that merit Greek letters – the brightest star, Praecipua, is catalogued as 46 Leonis Minoris according to John Flamsteed's numbering scheme (which gives the stars in a constellation increasing numbers with increasing right ascension). It is a typical orange giant of magnitude 3.8, about 98 light years from Earth and 22 times as luminous as the Sun in visible light. The most interesting star is not even visible to the naked eye – R Leonis Minoris is a long-period variable, a pulsating red giant similar to the famous Mira*. It varies in a period of 372 days, between magnitude 6.3 at its brightest, and magnitude 13 at its faintest.

Latin name:	Leo Minor
Latin genitive:	Leonis Minoris
Abbreviation:	LMi
Meaning:	Lesser Lion
Visibility:	Northern spring, southern autumn

Pyxis

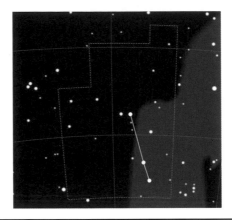

This unimpressive constellation is little more than a row of three stars occupying a small region of the sky near Puppis. It was invented by French astronomer Nicolas de Lacaille and represents a magnetic compass needle, so its location close to the remains of the great ship Argo is apt. Its most interesting object is a faint star that normally lies beyond the reach of small telescopes. T Pyxidis is a recurrent nova that flares up every couple of decades (most recently in 1996) to reach binocular or even naked-eye visibility. It is a binary system in which a white dwarf is pulling material out of the atmosphere of a larger companion star and down on to itself. Compressed by the dwarf's strong gravity, the layers of material wrapped around it eventually become hot and dense enough to ignite and burn away in a burst of nuclear fusion.

Latin name:	Pyxis
Latin genitive:	Pyxidis
Abbreviation:	Pyx
Meaning:	Compass
Visibility:	Northern spring, southern autumn

Microscopium

A small and obscure constellation invented by Frenchman Nicolas Louis de Lacaille to fill a gap in the southern sky, Microscopium lies between the brighter stars of Sagittarius and brilliant Fomalhaut in Piscis Austrinus. Like many of Lacaille's constellations, it represents a scientific instrument – in this case a microscope – but its faint scattering of stars bears little resemblance to anything. Gamma (γ) and Epsilon (ε) are the brightest stars, with magnitude 4.9 Alpha (α) Mic roughly 0.2 magnitudes fainter than either of them. It is a yellow giant 380 light years away, with a companion of magnitude 10.6 visible through a small telescope. AU Mic is a much closer red dwarf of magnitude 8.6, just 32 light years away and very young. It was born less than 10 million years ago, and is still surrounded by a disk of gas and dust. Gaps in this disk may mark the orbits of a newly formed planetary system.

Latin name:	Microscopium
Latin genitive:	Microscopii
Abbreviation:	Mic
Meaning:	Microscope
Visibility:	Northern summer, southern winter

Apus

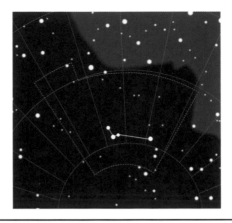

A small and shapeless constellation of the far southern sky, Apus was invented by Dutch trader/navigators Pieter Dirkszoon Keyser and Frederick de Houtman, who travelled in the East Indies in the late 1500s and named the constellation after the birds of paradise they encountered during their voyages Along with Grus and Phoenix, Apus forms a group known as the 'southern birds'. Delta (δ) Apodis is a double star consisting of a red giant of magnitude 4.7 and an orange giant of magnitude 5.3. Both are around 700 light years from Earth, but astronomers are unsure whether they are a chance alignment of two stars widely separated in space, or a true binary star. Theta (ϑ) Apodis is another red giant 330 light years away, classified as a semi-regular variable – it fluctuates between magnitudes 4.8 and 6.1 in a series of complex overlapping oscillations, dominated by a 5.9-day cycle.

Latin name:	Apus
Latin genitive:	Apodis
Abbreviation:	Aps
Meaning:	Bird of Paradise
Visibility:	Southern hemisphere circumpolar

Lacerta

This small but distinctive chain of stars was catalogued as a separate constellation by Polish astronomer Johannes Hevelius in 1787. Its zigzag pattern represents a small running lizard. Alpha (α) Lacertae is a blue-white star of magnitude 3.8, 102 light years from Earth and roughly 25 times as luminous as the Sun in visible light. Beta (β) Lac, which marks the lizard's nose, is more distant at around 170 light years. It is a yellow giant, and its magnitude of 4.4 indicates a luminosity equivalent to 40 Suns. The constellation straddles a bright patch of the Milky Way, and has played host to several novae, although none have so far put in a repeat performance in historical times. Its most famous object is probably BL Lacertae* – once mistaken for a bizarre variable star, but now known to be the prototype for an unusual class of active galaxy.

Latin name:	Lacerta
Latin genitive:	Lacertae
Abbreviation:	Lac
Meaning:	Lizard
Visibility:	Northern autumn, southern spring

179

Delphinus

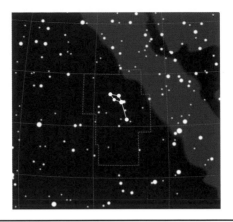

This small but distinctive constellation, lying just to the southwest of the Square of Pegasus, does bear a passing resemblance to the leaping dolphin it represents. In Greek myth, it commemorates a dolphin sent by the sea god Poseidon to rescue the celebrated bard Arion from drowning in a shipwreck. The unusual names of Alpha (α) and Beta (β) Delphini – Sualocin and Rotanev – are an astronomical joke, a backward spelling of Nicolaus Venator, which is itself a Latinized version of Niccolo Cacciatore, an Italian astronomer who worked at Palermo Observatory in the early 1800s. Both these stars are binary systems, though each is too tightly bound to be separable through an amateur telescope. Gamma (γ) Delphini, in contrast, is a well-separated double visible in good binoculars or a small telescope. Its brighter magnitude 4.1 star is yellow-orange, while the fainter companion of magnitude 5.1 is white, but can appear blue or green in contrast to its neighbour.

Latin name:	Delphinus
Latin genitive:	Delphini
Abbreviation:	Del
Meaning:	Dolphin
Visibility:	Northern summer, southern winter

Corvus

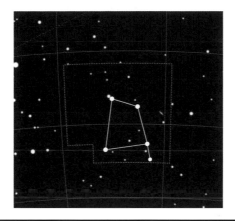

A small skewed rectangle of stars southwest of Spica in Virgo represents a crow, grasping at the great water snake Hydra in a legend that also takes in the neighbouring constellation Crater. Although none of its stars is particularly bright (the brightest is blue-white Gamma (γ) Corvi at magnitude 2.6), the constellation is still quite easy to spot. Delta (δ) Crv is an interesting binary star also known as Algorab (from the Arabic word for 'raven'). The yellow-white primary star, roughly 88 light years away, has a companion of magnitude 8.5 visible through a small telescope. This secondary star is often reported to be purple, although according to precise measurements of its light, its colour is really orange. The unusual apparent colour must arise through contrast with the brighter primary. Close to the border with Crater lies a faint but fascinating pair of galaxies, NGC 4038 and 4039, known as the Antennae*.

Latin name:	Corvus
Latin genitive:	Corvi
Abbreviation:	Crv
Meaning:	Crow
Visibility:	Northern spring, southern autumn

Canis Minor

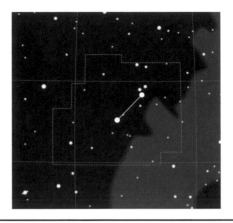

The smaller of Orion's two hunting dogs is a compact constellation to the east of the great hunter himself, easily overlooked except for the brightness of its primary star. Procyon's name comes from a Greek phrase meaning 'before the dog' – itself a reference to Procyon's tendency to rise shortly before the brilliant 'Dog Star' Sirius. Surprisingly, both stars have a lot in common: Procyon, like Sirius, is a near neighbour of the Sun, just 11.4 light years away. It weighs as much as 1.4 Suns, and pumps out seven times as much energy, which suggests it may have recently started down the path towards becoming a brilliant orange or red giant. Like the Dog Star itself, Procyon also has a small white dwarf companion – the remains of a more massive star that lived and died more rapidly than Procyon itself, and which long ago shrugged away its outer layers in a planetary nebula that has since faded and dispersed.

Latin name:	Canis Minor
Latin genitive:	Canis Minoris
Abbreviation:	CMi
Meaning:	Lesser Dog
Visibility:	Northern winter, southern summer

Dorado

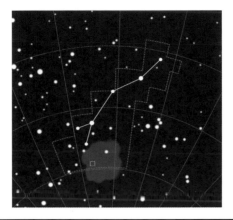

This constellation represents a tropical fish of some sort . Although its name is often translated as 'the Goldfish' or 'the Swordfish', the Dutch navigators Keyser and de Houtman, who invented it, probably intended it to depict the South Pacific fish known as a mahi-mahi. Although this chain of far southern stars is not particularly bright, it is easy to find because it plays host to the unmistakeable Large Magellanic Cloud* (LMC) – one of our galaxy's satellite systems. The LMC is easy to spot in dark skies, looking like a detached portion of the Milky Way itself. Binoculars or a small telescope will reveal rich starfields and features such as the spectacular Tarantula Nebula*. The constellation's most interesting star is Beta (β) Doradus, one of the brightest Cepheid variables in the sky. It varies between magnitudes 3.5 and 4.1 in a cycle that repeats every 9.9 days.

Latin name:	Dorado
Latin genitive:	Doradus
Abbreviation:	Dor
Meaning:	Goldfish
Visibility:	Southern hemisphere circumpolar

Corona Borealis

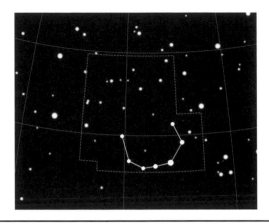

This small constellation is easy to identify because its stars, though only middling in brightness, form a distinctive curved coronet in the sky to the east of Boötes. They supposedly represent the crown worn by Ariadne at her wedding to Bacchus, the god of wine and festivities. Corona Borealis contains two faint but famous stars that are mirror-images of one another. R CrB* experiences sudden and catastrophic drops in brightness, while T CrB is a 'recurrent nova', consisting of a white dwarf in orbit around a bloated red giant, 1800 light years away. The dwarf star is pulling material from the giant's atmosphere on to itself, triggering periodic explosions on its surface (most recently in 1866 and 1946). Normally, the star hovers around magnitude 11, with nearly all of its light coming from the red giant, but in a nova event, the dwarf can brighten to reach magnitude 2.0 before slowly fading away – small wonder it is known as the 'Blaze Star'.

Latin name:	Corona Borealis
Latin genitive:	Coronae Borealis
Abbreviation:	CrB
Meaning:	Northern Crown
Visibility:	Northern summer, southern winter

Norma

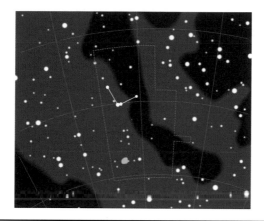

This small and obscure constellation is another invention of French astronomer Nicolas Louis de Lacaille, who surveyed the southern skies from South Africa in the 1750s. Its dim right-angle of stars represents a surveyor's level – a measuring tool that Lacaille placed close to Circinus, the compasses, in the heavens. Gamma (γ) Normae is the brightest star – an optical double that can usually be resolved with the naked eye. Its brighter star is a yellow giant of magnitude 4.0, 127 light years away, while its fainter component at magnitude 5.0, is in fact a far more luminous yellow supergiant 1500 light years from Earth. NGC 6087 is an attractive open cluster, just visible to the naked eye and 2700 light years away. Binoculars or a small telescope reveal about 40 stars, dominated by another bright yellow supergiant, S Normae. This star, a variable similar to Delta Cephei* oscillates between magnitudes 6.1 and 6.8 in a 9.8-day cycle.

Latin name:	Norma
Latin genitive:	Normae
Abbreviation:	Nor
Meaning:	Square
Visibility:	Southern hemisphere circumpolar

Mensa

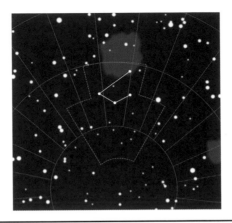

This small quadrangle of stars was first noticed by French astronomer Nicolas Louis de Lacaille during his observations from Cape Town in the 1750s. It seems that the pattern's relationship to the nearby Large Magellanic Cloud* reminded him of the clouds hanging over the nearby flat-topped Table Mountain, since he decided to name the constellation after it. With no stars brighter than magnitude 5, Mensa is the faintest constellation in the sky, and aside from a slight overspill of the LMC from neighbouring Dorado, it has no deep-sky objects of interest either. Alpha (α) Mensae is of some interest, though, as a comparison with our Sun – it is a relatively nearby yellow dwarf shining at magnitude 5.1 across a distance of 33 light years. It has 90 per cent of the Sun's mass, and is roughly 80 per cent as luminous. Even though conditions seem favourable, searches for planets orbiting Alpha Mensae have so far drawn a blank.

Latin name:	Mensa
Latin genitive:	Mensae
Abbreviation:	Men
Meaning:	Table
Visibility:	Southern hemisphere circumpolar

Volans

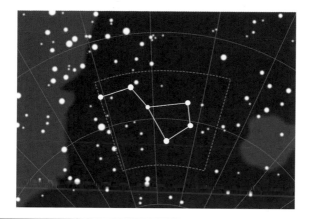

This small and obscure southern constellation represents the bizarre 'flying fish'. It was invented by Dutch navigators Frederick de Houtman and Pieter Dirkszoon Keyser, who explored the East Indies in the late sixteenth century and would doubtless have seen these creatures during their travels. Gamma (γ) Volantis is a beautiful double star with overall magnitude 3.2, whose two components have contrasting orange and white colours, revealed in a small or medium-sized telescope. The brighter star is an orange giant of magnitude 3.8, 147 light years from Earth, while the fainter one is a normal white star of magnitude 5.7. Epsilon (ε) Volantis is another double. Its primary star is blue-white, with the mass of about six Suns and a magnitude of 4.4. Its fainter but more widely separated companion is yellow-white, twice the mass of the Sun, and shines at magnitude 8.1 over a distance of 640 light years.

Latin name:	Volans
Latin genitive:	Volantis
Abbreviation:	Vol
Meaning:	Flying Fish
Visibility:	Southern hemisphere circumpolar

Musca

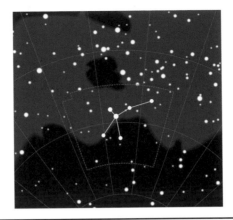

This small flurry of stars to the south of Crux was first noted by Dutch sailors Pieter Dirkszoon Keyser and Frederick de Houtman in the 1590s, and was originally Apis, the Bee. It was changed into Musca, the Fly, in the seventeenth century, perhaps to avoid confusion with another of Keyser and de Houtman's constellations, Apus, the Bird of Paradise. Its most interesting star is Theta (ϑ) Muscae – a double star easily split by a small telescope to reveal components of magnitudes 5.7 and 7.3, an impressive 10,000 light years away. The brighter component is a hot and brilliant blue star, but the fainter one is an unusual 'Wolf-Rayet' star – a stellar monster that started its life weighing as much as 50 Suns, but has developed such strong stellar winds that it has blown away its outer layers, leaving its super-hot inner material exposed to space. Elsewhere, NGC 4813 is an attractive globular cluster, easily located with binoculars.

Latin name:	Musca
Latin genitive:	Muscae
Abbreviation:	Mus
Meaning:	Fly
Visibility:	Southern hemisphere circumpolar

Triangulum

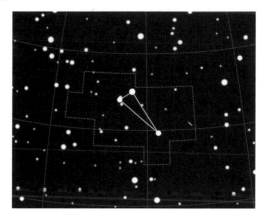

A narrow triangle of moderately bright stars drives a wedge between Perseus and Pegasus to the north of Aries. Although seemingly obscure, this stellar group was a constellation recognized by the ancient Greek astronomers – perhaps because of its resemblance to their capital letter delta (Δ). For a while, this was Triangulum Major, after the Polish astronomer Johannes Hevelius invented an even smaller Triangulum Minor (long since discarded). Triangulum's stars are unimpressive – the brightest, Beta Trianguli, shines at magnitude 3.0 and is a bright white star about 125 light years away. However, the constellation offers one outstanding sight – the Triangulum Galaxy*, or M33. This face-on spiral of stars, roughly three million light years away, is one of the closest major galaxies, and part of our own Local Group. With an apparent width equivalent to the full Moon, its light is so spread out that it can be difficult to find – look for it with binoculars or a low-powered eyepiece.

Latin name:	Triangulum
Latin genitive:	Trianguli
Abbreviation:	Tri
Meaning:	Triangle
Visibility:	Northern Autumn, southern Spring

Chamaeleon

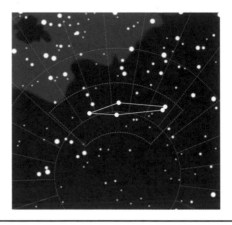

Chamaeleon is another of the faint constellations encircling the South Celestial Pole – an invention of Dutch explorers and traders Pieter Dirkszoon Keyser and Frederick de Houtman in the late 1500s, introduced to Europe by Petrus Plancius a few years later. Delta (δ) Chamaeleontis is a double star, thought to be a mere line-of-sight effect, although its stars appear to lie just a few light years apart. Binoculars will reveal the system as a blue-white star of magnitude 4.4, with a fainter companion that is itself a very tight binary system divisible only through the most powerful telescopes. Epsilon (ε) Cha is a close double for moderate-sized telescopes, consisting of twin white stars of magnitudes 5.4 and 6.0, 365 light years from Earth. NGC 3195, between Delta and Theta (ϑ) Cha, is the brightest planetary nebula of the far southern skies, 5500 light years away and visible through a small telescope.

Latin name:	Chamaleon
Latin genitive:	Chamaleontis
Abbreviation:	Cha
Meaning:	Chameleon
Visibility:	Southern hemisphere circumpolar

Corona Australis

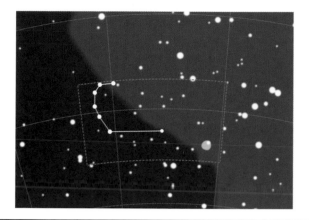

The southern crown is not quite such a perfect arc of stars as its northern equivalent, but it was still obvious enough to be recorded by the ancient Greek astronomers during its brief appearances above Mediterranean horizons. It lies south of the bulk of Sagittarius, and was mythologically associated with both the centaur (as Coronae Sagitarii) and the Greek god Bacchus (husband of Ariadne, wearer of the northern crown). Gamma (γ) Coronae Australis is an attractive double consisting of near-twin yellow-white stars with magnitues 4.8 and 5.1. Each has about 1.5 times the mass of the Sun, hence they burn brighter and have somewhat hotter surface temperatures. The two stars orbit each other in 120 years, and can currently be separated with small telescopes. NGC 6541, almost on the border of the Milky Way, is a globular cluster 22,000 light years away. Although just below naked eye visibility, it is easily spotted with binoculars.

Latin name:	Corona Australis
Latin genitive:	Coronae Australis
Abbreviation:	CrA
Meaning:	Southern Crown
Visibility:	Northern summer, southern winter

Caelum

One of the least impressive constellations in the sky, Caelum represents a chisel, and is little more than a line drawn between two fairly faint stars. Perhaps unsurprisingly, it is an invention of Nicolas Louis de Lacaille, the French surveyor of southern skies whose other contributions include the uninspiring Pyxis and Telescopium. The constellation has little to recommend it to an observer – its most interesting star is probably the variable R Caeli, a slowly pulsating red-giant variable that ranges between magnitudes 6.7 (easily visible in binoculars), and 13.7 (beyond most small telescopes) in a cycle lasting 400 days. Alpha (α) Caeli is the constellation's brightest star at magnitude 4.5, an average white main-sequence star 66 light years away. Slightly more interesting is Gamma (γ) Caeli, an orange giant 185 light years from Earth, with a close companion of magnitude 8.1 that can be separated in medium-sized telescopes.

Latin name:	Caelum
Latin genitive:	Caeli
Abbreviation:	Cae
Meaning:	Chisel
Visibility:	Northern winter, southern summer

Reticulum

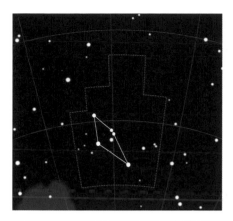

This small kite-shaped group of stars is another invention of Nicolas de Lacaille, the French astronomer who conducted the first systematic survey of southern skies from Cape Town in the 1750s. Although its name comes from the Latin for 'net', Lacaille actually intended it to represent a reticle – the cross-hair often inserted into telescope eyepieces to aid precise tracking or measurement. Alpha (α) Reticuli is a binary system 165 light years from Earth. Dominated by a yellow giant three times the mass of the Sun and shining at magnitude 3.4, its secondary star is a red dwarf of magnitude 12, visible only through medium-sized telescopes. Zeta (ζ) Reticuli is another binary system, just 39 light years from Earth and more easily separated. Indeed, the stars should be divisible with just the naked eye or binoculars. With magnitudes 5.2 and 5.5, the stars are near-identical yellow dwarfs very similar to the Sun.

Latin name:	Reticulum
Latin genitive:	Reticuli
Abbreviation:	Ret
Meaning:	Reticle
Visibility:	Northern winter, southern summer

Triangulum Australe

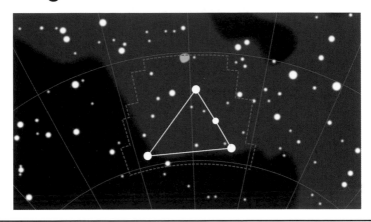

This broad, near-equilateral triangle of fairly bright stars lies to the east of the fainter, narrow triangle of Circinus and the bright stars Alpha and Beta Centauri. It is simply 'the Southern Triangle', first recorded by German astronomer Johann Bayer in his 1603 star atlas *Uranometra,* but probably introduced to Europe a few years earlier, in reports from Dutch navigator Pieter Dirkszoon Keyser. However, the constellation had long been established among Arab astronomers. Since they lived at lower latitudes, the pattern spent some time above their southern horizons, and the Arabs associated its stars with the three biblical patriarchs. The brightest star is Atria – a simple abbreviation of Alpha (α) Trianguli Australis. At magnitude 1.9, it is an orange giant 415 light years away, nearing the end of its life and shining by the nuclear fusion of helium in its core to form heavier elements.

Latin name:	Triangulum Australe
Latin genitive:	Trianguli Australis
Abbreviation:	TrA
Meaning:	Southern Triangle
Visibility:	Southern hemisphere circumpolar

Scutum

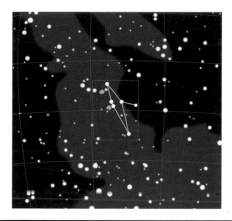

This constellation, invented by Polish astronomer Johannes Hevelius in 1684, represents a shield. Indeed, it was originally Scutum Sobieskii, after Hevelius's patron, the Polish king John III Sobieski (1629–96). Although its stars are not especially bright, Scutum lies in a dense area of the Milky Way. Delta (δ) Scuti is the prototype for a class of short-period variables, oscillating between magnitudes 4.7 and 4.8 in a few hours. Telescopes reveal it to be the central component of a triple star system. R Scuti is another variable, part of a rare class known as RV Tauri stars. It is a yellow supergiant that normally shines at magnitude 4.5 across 2500 light years, but dips in brightness every 71 days, alternating between 'deep' and 'shallow' minima of magnitudes 8.8 and 6.0. M11 is the 'Wild Duck' cluster, a huge and dense open cluster 6000 light years away, easily spotted through binoculars.

Latin name:	Scutum
Latin genitive:	Scuti
Abbreviation:	Sct
Meaning:	Shield
Visibility:	Northern summer, southern winter

Circinus

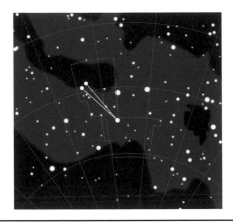

One of the many constellations added to the sky by Frenchman Nicolas de Lacaille during his observations from Capetown in the 1750s, Circinus represents a pair of drawing compasses used by surveyors and mapmakers. (It is not to be confused with Lacaille's Pyxis, which represents a magnetic compass.) Its stars, all of middling brightness, form a long thin triangle to the east of the bright stars Alpha and Beta Centauri. Lambda (λ) Circini is, in fact, a star cluster originally misclassified due to its naked-eye appearance as a fuzzy star. Today it is classified as NGC 6025, a fairly small open cluster 2700 light years away, best studied with binoculars. Theta (ϑ) Cir is a hot blue-white star, unpredictably variable between magnitudes 5.0 and 5.5. Alpha (α) Cir is a binary consisting of a magnitude 3.2 white star subject to small, rapid variations, and a fainter orange star of magnitude 8.6.

Latin name:	Circinus
Latin genitive:	Circini
Abbreviation:	Cir
Meaning:	Compasses
Visibility:	Southern hemisphere circumpolar

Sagitta

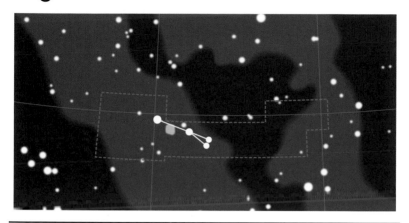

Surprisingly, this small arrow-like constellation of middle-ranking stars has nothing to do with the archer Sagittarius. It is one of Ptolemy's original 48 constellations, and apparently represents an arrow fired by Hercules in the direction of two birds: Cygnus the Swan, and Aquila the Eagle. Although compact, Sagitta contains several interesting objects. Gamma (γ) Sagittae, its brightest star, is one of the coolest stars visible to the naked eye – at magnitude 3.5, it is an extremely bloated red giant 275 light years away. M71, midway between Gamma and Delta Sagittae, is a globular cluster visible to binoculars, with an unusually loose and open structure revealed when it is viewed through a telescope. Astronomers used to argue over whether it was a true globular or just a dense open cluster. To its south lies WZ Sagittae, a recurrent nova that reaches binocular visibility during its outbursts, the last of which occurred in 2001.

Latin name:	Sagitta
Latin genitive:	Sagittae
Abbreviation:	Sge
Meaning:	Arrow
Visibility:	Northern summer, southern winter

Equuleus

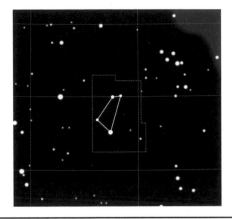

This skewed rectangle of stars southwest of Pegasus is supposed to represent a foal or small horse, and is often interpreted as Celeris, the offspring of the more famous winged horse. However, the pattern bears no resemblance either to a horse or to its larger neighbouring constellation. Alpha (α) Equulei is known as Kitalpha. At magnitude 3.9, it is 186 light years from Earth and is a spectroscopic binary – a yellow giant and a white main-sequence star locked in a 99-day orbit around one another, too close to be distinguished with even the most powerful telescope. Epsilon (ε) Equulei is an interesting triple star – small telescopes will show that the white magnitude 5.5 star has a yellow companion of magnitude 7.4. Larger instruments will show that the primary is, in fact, made up of two separate elements, with magnitudes 6.0 and 6.3.

Latin name:	Equuleus
Latin genitive:	Equulei
Abbreviation:	Equ
Meaning:	Foal
Visibility:	Northern summer, southern winter

Crux

The smallest constellation in the heavens is also one of the brightest – a compact and distinctive cross of four bright stars first noted by early European navigators in the southern hemisphere. Although bright, it is not completely unmistakcable, since a group of stars in nearby Carina form a slightly larger 'false cross' to the west. Alpha (α) Crucis, or Acrux, is a beautiful multiple star, easily split by a small telescope to reveal twin blue-white stars of magnitudes 1.3 and 1.7. The brighter of these stars is a spectroscopic binary in its own right. All three stars are likely to end their lives in supernova explosions. Gamma (γ) Crucis is a red giant of magnitude 1.6, contrasting obviously with the constellation's other blue-white stars. The southeast quadrant of the cross is marked by a dark dust cloud, the Coal Sack, which blocks the view of the Milky Way beyond. Nearby lies the beautiful Jewel Box cluster *.

Latin name:	Crux
Latin genitive:	Crucis
Abbreviation:	Cru
Meaning:	Southern Cross
Visibility:	Southern hemisphere circumpolar

The Sun

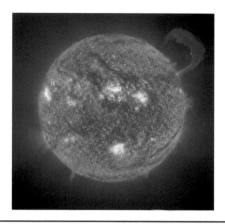

When considered among the mass of stars in the Milky Way galaxy, the Sun is nothing special – a small-to-average yellow star, typical of the metal-rich 'population I' stars that make up the majority of the galaxy. In fact, its most unusual feature is that it is solitary, since most stars are gravitationally bound together in binary or multiple star systems. Like most stars, the Sun lies on the so-called 'main sequence' of stellar evolution – a period in which a star's mass and luminosity (total energy output) follow a simple proportional relationship, while the star is shining by nuclear fusion of hydrogen into helium at its core. It has obeyed this rule, which places it neatly on a band between faint cool red stars and bright hot blue ones, for five billion years, and will shine steadily for another five billion to come.

Formal designation:	Sol
Right ascension:	N/A
Declination:	N/A
Constellation:	Centaurus
Object type:	G2V main-sequence yellow star
Magnitude:	-26.7
Distance:	1AU

Proxima Centauri

The closest star to Earth is so faint that its significance was not discovered until 1915. Proxima Centauri is an outlying member of the Alpha (α) Centauri* triple star system, lying 0.15 light years closer to the Sun and orbiting the central pair of stars in several hundred thousand years. With a mass just 20 per cent of the Sun's, its core is much less hot and dense, meaning that its nuclear reactions proceed more slowly. It releases far less energy and has a far fainter, cooler surface. Proxima is a red-dwarf star, and one advantage of its sedate lifestyle is that it will continue to shine for many, many billions of years into the future. Red dwarfs may make up the majority of stars in the galaxy, but they are so faint that they become extremely difficult to detect beyond the immediate neighborhood of our solar system.

Formal designation:	α Centauri C
Right ascension:	14h 30m
Declination:	-62° 41'
Constellation:	Centaurus
Object type:	Main-sequence red-dwarf star
Magnitude:	11.1
Distance:	4.22 light years

Alpha Centauri

The closest bright star, and the third brightest star in the entire sky, is Alpha (α) Centauri, sometimes known as Rigil Kentaurus (meaning 'foot of the Centaur'). It is a multiple star, consisting of a yellow primary star 10 per cent more massive and 50 per cent more luminous than the Sun, and shining at magnitude 0.0, orbiting with an orange star slightly less massive and luminous than our star at magnitude 1.4. The two stars circle each other every 80 years, and are in turn circled over a longer period by the smaller and fainter Proxima Centauri*. The entire system is thought to be about eight billion years old – three billion years older than the Sun – and the bright primary star, Rigel Kentauus A, is probably on the verge of exhausting its supply of hydrogen fuel, and swelling to become a huge, far more luminous giant.

Formal designation:	α Centauri (A and B)
Right ascension:	14h 40m
Declination:	-60° 50′
Constellation:	Centaurus
Object type:	Multiple star (main-sequence yellow and orange stars)
Magnitude:	-0.1
Distance:	4.36 light years

Barnard's Star

Although it is the fourth closest star to the Sun (the other three being the various members of the Alpha Centauri system), Barnard's Star was discovered by American astronomer E. E. Barnard only in 1916. It is a faint red dwarf similar to Proxima Centauri, with a mass around 20 per cent of the Sun's, and 1/300th of its luminosity. It has the largest 'proper motion' (movement through the sky relative to the solar system) of any star in the sky – large enough that it shifts its position in the sky by the width of a full Moon (half a degree) in just 180 years. Despite its feeble energy output, Barnard's Star has a strong magnetic field, like other red dwarfs, and its surface is subject to unpredictable flares large enough to alter its brightness. We still do not understand how such small stars create such mighty eruptions.

Formal designation:	V2500 Ophiuchi
Right ascension:	17h 58m
Declination:	+4° 42′
Constellation:	Ophiuchus
Object type:	Main-sequence red-dwarf star
Magnitude:	9.54
Distance:	5.9 light years

Wolf 359

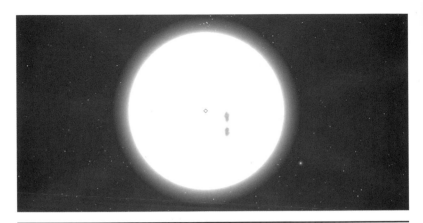

This red dwarf star in Leo is even fainter than the other nearby dwarf stars – at magnitude 13.5, it is visible only through a medium-sized telescope. The star was discovered in 1918 by German astronomer Max Wolf, using astrophotography – the comparison of photographic plates taken months apart showed that the star had shifted relative to the more distant background stars. This movement is called parallax, and is created by our changing view of the stars as the Earth orbits around the Sun. The star itself has about 10 per cent of the Sun's mass and 0.002 percent of its luminosity. Like the other nearby red dwarf stars, it is subject to solar flares that can cause it to double in brightness. Hubble Space Telescope observations found that it typically suffers from a dozen or more of these eruptions in an hour.

Formal designation:	V2500 Ophiuchi
Right ascension:	10h 56m
Declination:	+7° 01′
Constellation:	Leo
Object type:	Main-sequence red-dwarf star
Magnitude:	13.5
Distance:	7.8 light years

UV Ceti (Luyten 726-8 A & B)

Technically, UV Ceti is the dimmer component of this red dwarf binary whose proper name is Luyten 726-8, but UV Ceti is often used as shorthand for the entire system. This pair of faint dwarf stars lie 8.7 light years from the Sun and shine at magnitudes 12.5 and 13.0. They orbit each other in 26.5 years, and were discovered only in 1948. UV Ceti is subject to impressive stellar flares, which can increase its brightness by a factor of 5. The brighter component, technically BL Ceti, is a less extreme example of these 'flare stars'. The sheer number of dwarf stars in our solar neighbourhood suggests that they are widely scattered throughout the galaxy, and probably form the vast majority of stars. It is only their extreme faintness that prevents us detecting them in large numbers across greater distances.

Formal designation:	Lutyen 726-8 A & B
Right ascension:	1h 39m
Declination:	-17° 57'
Constellation:	Cetus
Object type:	Binary star (main-sequence red-dwarf stars)
Magnitude:	12.5
Distance:	8.6 light years

Sirius

The famous 'Dog Star' is only the brightest in the sky because of its proximity to Earth. Just 8.6 light years away, it is an average star with about twice the mass of the Sun, pumping out about 25 times as much energy. Its hotter surface temperature of 9600°C (17,300°F) results from the combination of its higher energy output with a diameter 1.7 times that of the Sun, and makes the star shine with a pure blazing whiteness. Like most stars, Sirius is not alone in space – it is a binary system, with a fainter secondary that was first observed in 1862. Sirius B was the first 'white-dwarf' star identified – it is the burnt-out core of a star that originally had more mass than Sirius A, and so rushed through its life far more quickly. Unfortunately, at magnitude 8.5, Sirius B is easily lost in the glare from its bright companion.

Formal designation:	a Canis Majoris
Right ascension:	6h 45m
Declination:	-16° 43'
Constellation:	Canis Major
Object type:	Binary star (main-sequence white star and white dwarf)
Magnitude:	-1.46
Distance:	8.6 light years

Horsehead Nebula

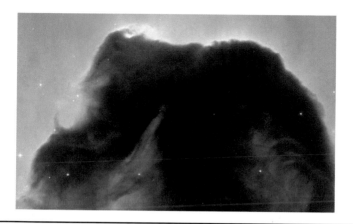

Between the stars lie huge clouds of gas and dust that give rise to new stellar generations. Some of these are luminous in their own right, others reflect light from the stars around them, but many are cold, dark and opaque, producing no visible light and very little radiation. In general, these 'dark nebulae' can be seen only where they are silhouetted against a brighter background – either a dense star field or a more remote nebula. The most famous of these celestial silhouettes is the Horsehead Nebula, near Alnitak in the belt of Orion. Known as Barnard 33, this dark dust cloud, about 3.5 light years across and 1500 light years away, is thrown into contrast against the curtains of light in IC 434, a vast cloud of molecular hydrogen that glows as its atoms are excited by ultraviolet radiation from Alnitak.

Formal designation:	Barnard 33/IC 434
Right ascension:	05h 41m
Declination:	-02° 28'
Constellation:	Orion
Object type:	Dark nebula
Magnitude:	N/A
Distance:	1500 light years

Orion Nebula

The brightest and most easily seen nebula in the entire sky, the Orion Nebula M42 lies within the celestial hunter's sword. A noticeably 'smudged' star to the naked eye, binoculars and telescopes reveal more detail in its flower-like shape. The nebula extends across an entire degree of sky – it is about 1500 light years away (part of the same complex as IC 404 and the Horsehead Nebula) and 15 light years across. Smaller and fainter nebulae extend above and below it. M42 is an emission nebula, glowing as its gas is excited by ultraviolet radiation from the stars forming at its centre. Most prominent of these are the newborn, hot blue-white stars of Theta Orionis, a quadruple star nicknamed the Trapezium. Binoculars show it as a double star, while a small telescope will split each element into a 'double double'.

Formal designation:	M42/NGC 1976
Right ascension:	5h 35m
Declination:	-5° 27'
Constellation:	Orion
Object type:	Emission nebula
Magnitude:	4.0
Distance:	1600 light years

Trifid Nebula

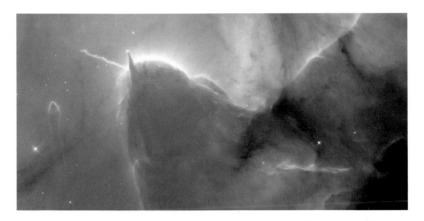

This nebula in Sagittarius combines aspects of emission, reflection and dark nebulae to produce a colourful star-forming cloud segmented into three by dark dust lanes. It was catalogued by Messier in 1764, and William Herschel first noted its triple-lobed appearance two decades later. At its centre, a young star cluster excites hydrogen gas in the surrounding nebula to emit a pinkish glow. A blue reflection nebula surrounds the central emission region, and the glowing clouds silhouette Barnard 85, a dark dust cloud. Most of the nebula's emission comes from a triple star (with a brightest component of magnitude 7.6) on the western edge of its central cluster. All the stars in this system are extremely hot and blue, and so emit most of their energy as ultraviolet radiation that interacts with the surrounding hydrogen.

Formal designation:	M20/NGC 6514
Right ascension:	18h 03m
Declination:	-23° 02'
Constellation:	Sagittarius
Object type:	Complex nebula
Magnitude:	9.0
Distance:	5200 light years

Keyhole Nebula

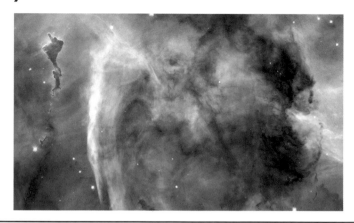

This dark column in the Carina Nebula marks an evaporating column of dust and gas in which stars are forming. Its silhouette punches a hole out of the background glow, and inspired John Herschel to name it the Keyhole while observing it from South Africa in the 1830s. The circular region at the top of the keyhole is about seven light years across, but as with many star-forming pillars, it is rapidly being eroded and broken into smaller fragments by the radiation from nearby young, hot stars. The entire Carina Nebula is a rich stellar nursery that has given birth to some of the most massive known stars, such as Eta Carinae*, and their fierce light strangles their siblings at birth, or at least stunts their growth. As with the Trifid Nebula, the Carina Nebula is a complex mix of emission, reflection and dark nebulae.

Formal designation:	NGC 3372
Right ascension:	10h 44m
Declination:	-59° 52′
Constellation:	Carina
Object type:	Complex nebula
Magnitude:	1.0
Distance:	8000 light years

Eagle Nebula

This huge but faint star-forming nebula surrounds the young open cluster M16 in Serpens Cauda. Only the largest amateur telescopes can view it directly, but long-exposure photographs show that it forms a glowing 'cavern' in the sky – a hollow shell of gas illuminated by the light of the young hot stars at its heart. Near the centre lie the finger-like projections known as the Pillars of Creation (a title coined after the Hubble Space Telescope photographed the area in detail). Each pillar is several light years long, an opaque column of gas and dust in which new stars and solar systems are forming. Along the edges, small dark globules indicate protostars that have broken free of the pillars, which are constantly being eroded and blown away by the radiation of earlier generations of stars.

Formal designation:	M16/NGC 6611
Right ascension:	18h 19m
Declination:	-13° 47'
Constellation:	Serpens
Object type:	Open star cluster and complex nebula
Magnitude:	6.4
Distance:	7000 light years

Thackeray's Globules

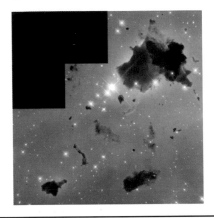

Dark knots of gas called Bok globules (after their discoverer, Dutch-American astronomer Bart Bok) are the closest we come to seeing the birth of a star. Visible when silhouetted against emission nebulae, these are isolated regions of dust-rich material, cut off from larger star-forming pillars but held together by the gravity of the stars forming within them. Thackeray's Globules, identified in 1950 by South African astronomer A. D. Thackeray, lie in the constellation of Centaurus, almost 6000 light years away. The Australia-shaped mass at the centre of this image, captured by Hubble Space Telescope, consists of two globules at slightly different distances, overlapping because they are on the same line of sight from Earth. Each is about a light year across, and together they contain the mass of about 15 Suns.

Formal designation:	IC 2944
Right ascension:	11h 38m
Declination:	-63° 20'
Constellation:	Centaurus
Object type:	Dark globules in emission nebula
Magnitude:	N/A
Distance:	5900 light years

T Tauri

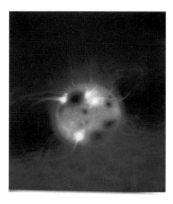

This unpredictable variable has a magnitude of between 9.6 and about 14. A young star, it is less than a million years old and suffering from birth pangs as it grows denser and hotter. Eventually, conditions in its core will become extreme enough to trigger the fusion of hydrogen into helium, and the energy released by the first sparks of this process will trigger a chain reaction that fires the heart of the star into life and stabilizes it. At this point, T Tauri will take up a long-term position on the 'main sequence' of stellar evolution, with a surface temperature and energy output governed by its mass. But for the moment, it shines less predictably through the energy released by its gravitational contraction, and the less demanding process of fusing deuterium (a rare type of 'heavy' hydrogen atom).

Formal designation:	T Tauri
Right ascension:	4h 22m
Declination:	+19° 32'
Constellation:	Taurus
Object type:	Pre-main sequence variable star
Magnitude:	9.6 (var)
Distance:	575 light years

HH 32

Even after a star's core has fired into hydrogen-fusing life and the star has begun to shine properly, it will not settle down, since it is usually still surrounded by a spinning disk of gas and dust. Beyond a certain distance, this material may remain in a stable orbit and begin to form a planetary system. Any closer, and it will fall in towards the star. The gas can form a dense 'accretion disc', spinning faster towards the centre, and where it contacts the star, material can be flung off in narrow jets from its poles. These jets are almost invisible as they shoot across space, but when they encounter clouds of interstellar gas, they can cause them to glow, creating an emission nebula called a Herbig-Haro object. HH 32 is among the brightest of these objects, about 1000 light years away in the constellation Aquila.

Formal designation:	HH 32
Right ascension:	19h 21m
Declination:	+11° 01'
Constellation:	Aquila
Object type:	Herbig Haro object
Magnitude:	7.9 (central star)
Distance:	960 light years

The Jewel Box

A beautiful star cluster, the Jewel Box, or Kappa Crucis cluster, was named by Sir John Herschel. It lies in the sky's smallest constellation, Crux, and makes a striking contrast to the nearby dark nebula known as the Coalsack. The brightest star, Kappa Crucis, is hot and blue-white at magnitude 5.9, but binoculars or a small telescope reveal a colourful range of stars, including a red giant of magnitude 7.6. Astronomers estimate the age of open clusters like the Jewel Box from how far their more massive stars have evolved. Since many of this cluster's massive, short-lived blue stars are still in the main sequence phase of their evolution, it is probably very young – around seven million years old. Based on this figure, its lone red giant must be a true stellar heavyweight to have outpaced its siblings in the evolutionary race.

Formal designation:	NGC 4755
Right ascension:	12h 54m
Declination:	-60° 20′
Constellation:	Crux
Object type:	Open star cluster
Magnitude:	4.2
Distance:	7600 light years

The Pleiades

Arguably the best-known star cluster in the sky, the Pleiades, or Seven Sisters, form a distinctive hook-shaped blob on the back of the charging bull Taurus. Naked-eye observers usually pick out just six stars, but binoculars or a telescope with a wide field of view reveal dozens more. The brightest star is Alcyone, at magnitude 2.9, and another interesting member is Pleione, a rapidly spinning star similar to Beta (β) Cassiopeiae. The cluster lies just 440 light years away, and has an overall magnitude of 1.6, although its light is spread out over a large area of sky. A dark night and sharp eyes may reveal tenuous traces of gas still surrounding some of the stars – the Pleiades are a relatively young cluster at just 50 million years old. This youthfulness means the cluster is dominated by short-lived, blue-white stars.

Formal designation:	M45
Right ascension:	3h 47m
Declination:	+24°
Constellation:	Taurus
Object type:	Open star cluster
Magnitude:	1.6
Distance:	440 light years

The Hyades

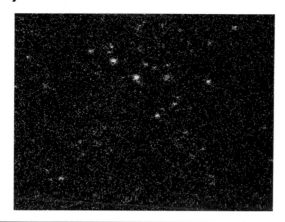

This V-shaped open cluster, which forms the face of Taurus, the Bull, is the closest noticeable star cluster to Earth, about 150 light years away. It bears the name of a group of nymphs from Greek myth, sisters of the Pleiades. The Ursa Major Moving Group is closer, but fills such a large area of sky that it is hard to appreciate as a cluster. With about 200 stars scattered across several degrees of the sky, the Hyades are impossible to miss, although their brightest member, orange-red Aldebaran*, is in fact a closer giant that lies in their direction. They occupy about 80 light years of space, and are considerably older than the Pleiades – roughly 800 million years old. This means that the heaviest, brightest stars have already burned out, and the cluster is now dominated by middle-ranking stars of various colours.

Formal designation:	Melotte 25
Right ascension:	4h 27m
Declination:	+16°
Constellation:	Taurus
Object type:	Open star cluster
Magnitude:	N/A
Distance:	150 light years

217

The Beehive Cluster

This open cluster is one of the finest visible to the naked eye from Earth. It was first recorded in the third century BC, and is also known as Praesepe (from the Latin for 'manger', since its stars were seen as a trough full of hay), but the Italian astronomer Galileo Galilei was one of the first to distinguish its stars when he turned his primitive telescope towards the cluster in the early 1600s. Modern binoculars or a small telescope will reveal that the Beehive contains about 200 stars in a region about three times the width of the Full Moon, with Epsilon Cancri its brightest individual member at magnitude 6.3. The cluster is about 550 light years from Earth, but measurements of its motion suggest it originated in the same place as the much closer Hyades cluster in Taurus, about 800 million years ago.

Formal designation:	M44/NGC 2632
Right ascension:	8h 40m
Declination:	+19° 59'
Constellation:	Cancer
Object type:	Open star cluster
Magnitude:	3.7
Distance:	580 light years

Spica

The brightest star in Virgo and one of the brightest in the sky, Spica is actually a binary system, consisting of two near-identical blue-white stars that orbit each other in just four days. The twin stars give themselves away through their effect on each other's spectral lines, which shift this way and that as the stars move back and forth from our point of view thanks to the Doppler effect, and because they distort each other's shapes to make the system slightly variable. The system lies 260 light years away, which means that the two stars must together produce the same amount of visible light as 2100 Suns. However, because their surfaces are so hot, at around 20,000°C (36,000°F), they release five times as much energy as invisible ultraviolet radiation – many hot stars look deceptively faint for this reason.

Formal designation:	α Virginis
Right ascension:	13h 25m
Declination:	-11° 10′
Constellation:	Virgo
Object type:	Multiple star (blue main-sequence stars)
Magnitude:	1.0
Distance:	260 light years

Vega

A typical white star, Vega is one of the most prominent in the sky, the brightest star in Lyra and one corner of the 'summer triangle' of northern skies. With the mass of 2.3 Suns, Vega shines by the same nuclear fusion process that drives the Sun, transforming hydrogen nuclei into helium. However, while stars like the Sun build up their helium atoms bit by bit in the proton-proton chain, more massive stars with higher internal pressures can use a faster process called the CNO cycle, in which carbon atoms in the star's core act as a catalyst to help helium form more easily. The result is an increase in the brilliance of stars more massive than the Sun – Vega, for example, produces 40 times the energy of the Sun. Its radius is roughly 2.7 times solar, so this energy heats its surface to a white-hot 9200°C (16,500°F).

Formal designation:	α Lyrae
Right ascension:	18h 37m
Declination:	+38° 47′
Constellation:	Lyra
Object type:	White main-sequence star
Magnitude:	0.0
Distance:	25 light years

Epsilon Eridani

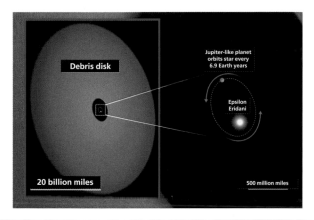

Debris disk

20 billion miles

Jupiter-like planet
orbits star every
6.9 Earth years

Epsilon
Eridani

500 million miles

A yellow star of magnitude 3.4 shining in the northern reaches of the river constellation Eridanus, Epsilon (ε) Eridani is a fairly yellow star. It has 80 per cent of the Sun's mass and radiates about one third of its light. It is visible to the naked eye only because it lies on our cosmic doorstep at a distance of 10.5 light years. Epsilon is shining by the same processes that power the Sun – the fusion of hydrogen into helium in a chain reaction called the proton-proton chain. This is a comparatively slow and undemanding fusion process, so the star will spend many billions of years on the main sequence, even though it has less fuel to burn than larger stars. Like many sunlike stars, Epsilon Eridani has a planetary system of sorts – a giant planet at least as big as Jupiter, and a dust ring further out from the star.

Formal designation:	ε Eridani
Right ascension:	3h 33m
Declination:	-09° 28'
Constellation:	Eridanus
Object type:	Yellow main-sequence star with extrasolar planetary system
Magnitude:	3.4
Distance:	10.5 light years

221

61 Cygni

This inconspicuous star in Cygnus, apparently of magnitude 5.0, is transformed by binoculars into a pair of near twin orange stars with magnitudes 5.2 and 6.0. The stars have masses equivalent to 0.5 and 0.4 Suns respectively, and luminosities of nine per cent and four per cent solar – a clear indication of how critical a star's mass is to its brightness. They are visible to the naked eye only because 61 Cygni is one of the closest stars to the Sun. In fact, it was the first star to have its distance directly measured, by German astronomer William Friedrich Bessel. In 1838, he calculated this from its parallax – the amount that its position apparently shifts throughout the year as our point of view moves from one side of Earth's orbit to another. His measurement of 10.4 light years was within 10 per cent of the modern figure.

Formal designation:	61 Cygni
Right ascension:	21h 07m
Declination:	+38° 45′
Constellation:	Cygnus
Object type:	Binary star (main- sequence orange dwarf stars)
Magnitude:	5.0
Distance:	11.4 light years

Gliese 229

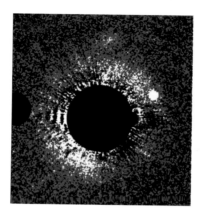

Low-mass stars spend their main-sequence lives as red dwarfs – feeble stars with low luminosities and cool surfaces. They shine by the same nuclear fusion reactions as the Sun – the proton-proton (p-p) chain version of hydrogen fusion – but these proceed slowly because the temperature and pressure of the star's core are lower. As a result, relatively nearby and bright red-dwarf stars such as Gliese 229 in Lepus are detectable only if they lie in our solar neighbourhood. The faintest stars have the mass of 0.08 Suns – below this, an object is not a star but a 'brown dwarf' shining by gravitational contraction and the deuterium reactions that fuel young stars (as on T Tauri *). Gliese 229's companion (Gliese 229b) was the first brown dwarf found – a 'failed star' the size of Jupiter but with a few dozen times its mass.

Formal designation:	Gliese 229 a and b
Right ascension:	6h 10m
Declination:	21° 52'
Constellation:	Lepus
Object type:	Main-sequence red-dwarf star with brown-dwarf companion
Magnitude:	8.1
Distance:	19 light years

Albireo

One of the most beautiful double stars in the sky, Albireo or Beta (β) Cygni is visible to the naked eye as a yellow star of magnitude 3.1, but steady binoculars or a small telescope reveal a blue-green companion of magnitude 5.1, forming a striking contrast with the brighter primary star. In fact, Albireo is a triple star – the brighter component, apparently yellow, is actually a close pairing of an orange giant of magnitude 3.3 and a hot blue star of magnitude 5.5. The inner pair are inseparable with traditional telescopes, and orbit each other in about 100 years, but the wider star is so far away that astronomers used to think the stars were a chance alignment rather than a true system linked by gravity. We now know, however, that the blue companion star completes its orbit in about 75,000 years.

Formal designation:	β Cygni
Right ascension:	19h 31m
Declination:	+27° 58'
Constellation:	Cygnus
Object type:	Binary star (orange giant and blue main-sequence star)
Magnitude:	3.1
Distance:	385 light years

Epsilon Lyrae

This beautiful multiple star is the sky's best-known 'double double'. Close to Vega in the constellation Lyra, Epsilon appears as a single fourth-magnitude star to naked eye observers, but binoculars (or even keen eyesight on a dark, clear night) will transform it into a pair of stars – northern Epsilon 1 with magnitude 4.7, and southern Epsilon 2 with magnitude 4.6. A small telescope splits these components again to reveal a pair of white stars within each. Epsilon 1's stars are uneven, with magnitudes 4.7 and 6.2, orbiting each other in about 1200 years. Epsilon 2 is a more even pair of magnitudes 5.1 and 5.5, locked in a 585-year orbital dance. The two pairs are 120 light years away, and separated from each other by roughly 0.16 light years, and must take several hundred thousand years to orbit one another.

Formal designation:	ϵ Lyrae
Right ascension:	18h 44m
Declination:	+39° 40'
Constellation:	Lyra
Object type:	Quadruple star (white main-sequence stars)
Magnitude:	4.6
Distance:	160 light years

Algol

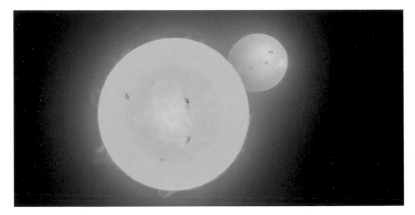

The 'winking demon' of Perseus is one of the most famous stars in the sky – an 'eclipsing binary' star that drops from magnitude 2.1 to 3.4 for 10 hours in every 69. The plunge in brightness is easily tracked with the naked eye, and was first recorded in 1670, although its name, which derives from Arabic, suggests that eastern astronomers may have recognized its behaviour far earlier. The British astronomer John Goodricke realized, in 1783, that the 'single' star must be an extremely close double, with the system aligned in the same plane as Earth so that the two stars pass behind and in front of one another, affecting the brightness seen from Earth. Today we know that Algol's primary star is blue-white and 100 times as luminous as the Sun, while the fainter star is a low-mass orange giant.

Formal designation:	β Persei
Right ascension:	3h 08m
Declination:	+40° 57′
Constellation:	Perseus
Object type:	Eclipsing binary system
Magnitude:	2.1 (var)
Distance:	93 light years

Epsilon Aurigae

The northernmost star in Auriga's triangle of 'Kids', Epsilon Augigae (also known as Almaaz) is an eclipsing variable, normally shining at magnitude 3.0. Its period is long, with an eclipse once every 27 years, but more strange is the nature of the eclipse itself – it lasts for about a year, during which the system's brightness drops to magnitude 3.8. This suggests that the object causing the eclipse is vast – perhaps 3 billion kilometres (1.8 billion miles) across – but it seems to be semi-transparent, since some light from the white supergiant primary star shines through it, without making a contribution to the system's overall light output. Most astronomers think it is a disc of gas and dust around a central object – perhaps a solar system being formed – but it is not clear why the star at the centre of it is essentially invisible.

Formal designation:	ε Aurigae
Right ascension:	05h 02m
Declination:	+43° 49'
Constellation:	Auriga
Object type:	Eclipsing binary system
Magnitude:	3.0 (var)
Distance:	2000 light years

Sheliak

The second brightest star in the northern constellation of Lyra, Sheliak or Beta Lyrae is a multiple star known as an interacting binary. A small telescope shows that the naked-eye star, white and usually of magnitude 3.5, has a bluish companion of magnitude 7.2. The brighter star is a spectroscopic binary in its own right, and dips to magnitude 4.3 every 12 days 22 hours as the larger star passes in front of the smaller one and blocks out its light. A secondary minimum, dipping to magnitude 3.8, happens 6.5 days later as the smaller star blocks out light from the larger one. In addition, the two stars are so close together that they distort each other's shapes into stretched 'ellipsoids'. As a result, the amount of visible surface they present to Earth is constantly changing, causing the system's brightness to vary.

Formal designation:	β Lyrae
Right ascension:	18h 50m
Declination:	+33° 22′
Constellation:	Lyra
Object type:	Eclipsing binary system (contact binary)
Magnitude:	3.5 (var)
Distance:	880 light years

Arcturus

When a star exhausts the hydrogen for nuclear fusion in its core, its structure alters and it stops obeying the 'main-sequence' relationship between luminosity, surface temperature and colour. Arcturus, a red-orange star in Boötes, is one such example. About 1.5 times the mass of the Sun, it ran out of core hydrogen some time ago. As radiation from the core faltered, the star's interior compressed until the layer of hydrogen-rich material around the core became dense enough to ignite in a spherical shell of nuclear fusion. This boosted the star's brightness, and the resulting increase in the outward force of radiation above the fusion shell made the star's upper layers balloon. The star's vastly increased surface now receives far less heating per unit area, so it has cooled, making Arcturus a 'red giant'.

Formal designation:	α Bootis
Right ascension:	14h 16m
Declination:	+19° 10'
Constellation:	Bootes
Object type:	Red-giant star
Magnitude:	0.0
Distance:	37 light years

Aldebaran

This orange-red giant star is the brightest in Taurus, marking the eye of the bull. Aldebaran is one of the closest giant stars to Earth, and with a mass of 2.5 times the Sun's, it has evolved beyond the red-giant phase. Even as hydrogen fusion continues in a spherical shell within a red giant, the exhausted helium-rich core is collapsing under its own weight. Eventually, conditions become so extreme that the helium residue ignites in its own fusion process, forming heavier elements such as carbon and oxygen. As the core restarts, the sudden change in internal conditions 'turns down' the hydrogen-burning shell so that luminosity drops and the outward pressure of radiation decreases. As the star shrinks under its own weight, its surface grows hotter, until the star has stabilized as a helium-burning giant.

Formal designation:	α Tauri
Right ascension:	4h 36m
Declination:	+16° 31'
Constellation:	Taurus
Object type:	Variable star (helium-burning orange giant)
Magnitude:	0.85 (var)
Distance:	65 light years

R Leonis

An easily tracked variable, R Leonis is a bright red star whose magnitude varies between 5.5 and 10 or 11 with a period of 312 days. It is similar to Mira* in Cetus, but is of interest because it represents a late stage in the evolution of stars like the Sun. With even the helium in its core exhausted, the star's central power source has been cut off, compressing the area around the core and intensifying the fusion of hydrogen into helium in a surrounding shell. A second shell of fusion, using the helium produced by the first to produce heavier elements, now begins, and the combined energy output from these two fusion shells is far greater than that from the core, so the star's outer layers swell once again and its surface cools to become an even larger red giant than before. In the process, it becomes unstable.

Formal designation:	R Leonis
Right ascension:	9h 48m
Declination:	+11° 26′
Constellation:	Leo
Object type:	Variable star (final-stage red giant)
Magnitude:	7.5 (var)
Distance:	390 light years

Ring Nebula

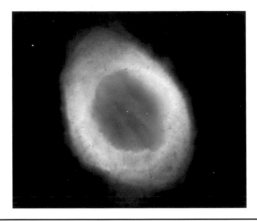

As a star like the Sun nears the end of its life, the pulsations rippling through it become more violent. The star may swell up, and then shrink suddenly, leaving behind expanding layers of gaslike smoke rings in space. The shells glow as they are excited by radiation from the shrinking star at their centre (whose surface grows hotter as each new layer is exposed). These stellar cast-offs are known as planetary nebulae (though they have nothing to do with planets save their spherical shape). A famous example is the Ring Nebula, M57, in Lyra. Astronomers used to think it was a spherical shell of gas with only its edges visible, but now suspect we are looking down on the pole of the star at its centre, and the ring is either a genuine 'torus' of material cast off around the star's centre, or even a cone of gas projected towards us.

Formal designation:	M57/NGC 6720
Right ascension:	18h 54m
Declination:	+33° 02'
Constellation:	Lyra
Object type:	Planetary nebula
Magnitude:	8.8
Distance:	2300 light years

Helix Nebula

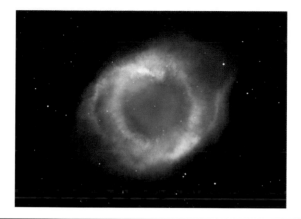

The closest planetary nebula to Earth, and the largest in the sky (with a diameter larger than the full Moon), the Helix Nebula in Aquarius is a challenging target because its light is spread over such a large area. Good binoculars may reveal a faint glowing disc of light, but long-exposure photographs are needed to show its true structure. The Helix is a roughly spherical shell of discarded gas, though with a concentrated ring around the equator of its central star, which we see at an angle. The structure is complicated by the fact that it contains two separate shells, expelled around 12,000 and 6000 years ago (indicating how short-lived planetary nebulae are in terms of stellar life cycles). In addition, the shape of the expanding shells is affected by interactions with the surrounding interstellar gas clouds.

Formal designation:	NGC 7293
Right ascension:	22h 30m
Declination:	-20° 48′
Constellation:	Aquarius
Object type:	Planetary nebula
Magnitude:	7.3
Distance:	450 light years

233

Cat's Eye Nebula

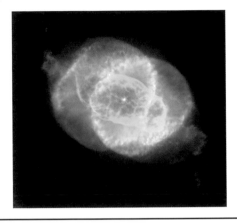

The Cat's Eye Nebula (NGC 6543) in Draco contains several overlapping shells and rings of gas and still poses several mysteries to astronomers. Astronomers put most of its structure down to the interaction of two processes – the central star is nearing the end of its life and casting off its outer layers as expected, but thanks to its high mass (originally five times that of the Sun) and hot surface temperature (about 80,000°C or 144,000°F), it is also producing a strong stellar wind that 'hollows out' the expanding shells of gas. Other structures, including jets that seem to emerge from the poles of the central star and trace spiral patterns through the nebula, lead astronomers to suspect that the central star is in fact binary, and may be surrounded by an 'accretion disk' of material being dragged from one star on to the other.

Formal designation:	NGC 6543
Right ascension:	17h 58m
Declination:	+66° 38'
Constellation:	Draco
Object type:	Planetary nebula
Magnitude:	8.1
Distance:	3600 light years

Hourglass Nebula

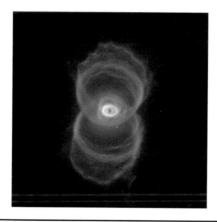

This nebula in the constellation Musca is nicknamed the 'Eye of God'. Tones in this false-colour image taken by the Hubble Space Telescope indicate different elements – green for hydrogen, red for nitrogen and blue for oxygen. The origins of the shape seem clear: shells of material shed from the central star's outer layers are 'blown up' by stellar winds, but are pinched around the system's equator by a slower-moving ring of denser gas. However, there are still questions about the nebula's finer structure. For example, what causes the detail within the outer shells of gas – interaction with the interstellar medium beyond them, or the result of polar jets from the central star disrupting the shells from within? Some of the Hourglass's features suggest that its star may be a binary similar to that in the Cat's Eye Nebula*.

Formal designation:	MyCn18
Right ascension:	13h 39m
Declination:	-67° 23'
Constellation:	Musca
Object type:	Planetary nebula
Magnitude:	13.0
Distance:	8000 light years

40 Eridani B

Omicron (o) Eridani, in the northern reaches of the constellation Eridanus, consists of two star systems on the same line of sight. The brighter star is 125 light years away and shines at magnitude 4.0, but the fainter one, also known as 40 Eridani, is just 16 light years from Earth, shining at magnitude 4.4. A small telescope reveals that 40 Eri is a true multiple star, with a magnitude 9.5 companion. 40 Eridani B, as this secondary star is known, is the easiest white dwarf to see from Earth, since it has no bright companion, unlike Sirius B and Procyon B. It is the shrunken but still hot and luminous core of a burnt-out star that cast off its outer layers in a planetary nebula. With a mass similar to the Sun, the white dwarf has enough gravity to hold on to a companion, a faint red dwarf of magnitude 11.2.

Formal designation:	o-2 Eridani
Right ascension:	4h 15m
Declination:	-07° 39′
Constellation:	Eridanus
Object type:	White-dwarf star in binary system
Magnitude:	9.5
Distance:	16.5 light years

DQ Herculis

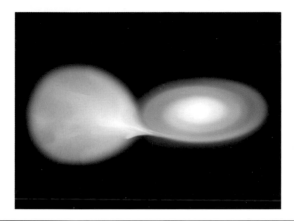

White dwarf stars in binary systems may affect their neighbours, giving rise to novae – stellar explosions such as wracked the usually unnoticeable star DQ Herculis in 1934. When the two elements of a close binary have different masses, one inevitably ages and evolves faster than the other, reaching its final state of a white dwarf while its companion is on the main sequence. The white dwarf's intense gravity may then drag material from the secondary's outer layers, pulling it down to create a compressed 'atmosphere' on its own surface, which grows denser and hotter over centuries. Eventually, atmospheric conditions become so intense that nuclear fusion reactions are triggered and the star flares to life, burning away its atmosphere and increasing in brightness by a factor of 10,000 or more.

Formal designation:	DQ Herculis
Right ascension:	18h 08m
Declination:	+45° 52′
Constellation:	Hercules
Object type:	Cataclysmic variable (nova)
Magnitude:	14.2
Distance:	1800 light years

RS Ophiuchi

Some novae explode more frequently than 'slow novae' like DQ Herculis. RS Ophiuchi has erupted five times in the past century (most recently in 2006), its brightness typically increasing by only a few hundred times in these outbursts. The star is a close binary, and the white dwarf is orbiting within the outer atmosphere of its companion, which has become a red giant. As the dwarf pulls material on to itself, it is thought to form a spiral accretion disc. This is inherently unstable, and occasionally collapses, dumping matter on to the white dwarf in a single event that triggers a nuclear outburst and a nova event. Although huge amounts of gas are dumped on to the star in a short period, the overall amount of material available for nuclear fusion is less than that provided by the slow build-up of a normal nova.

Formal designation:	RS Ophiuchi
Right ascension:	17h 50m
Declination:	-6° 42'
Constellation:	Ophiuchus
Object type:	Cataclysmic variable (recurrent nova)
Magnitude:	About 11.5 (when inactive)
Distance:	About 3500 light years

Supernova 1994D

Occasionally, a nova-like system of a white dwarf and a normal star will produce a stellar explosion that can outshine a galaxy. One such event was Supernova 1994D, which dominated the spiral galaxy NGC 4526 for a few weeks in 1994. 'Type Ia' supernovae happen because there is an upper mass limit for white dwarfs – any stellar remnant weighing more than 1.44 solar masses will collapse into a neutron star. When a white dwarf in a nova system reaches this limit, the material pouring on to it from its companion star may trigger a collapse and an outburst of energy almost equal to 'Type II' supernovae, in which a heavyweight star forms a neutron star or black hole. Because Type Ia supernovae are identical and release the same amount of energy, they are 'standard candles' for measuring intergalactic distances.

Formal designation:	Supernova 1994D
Right ascension:	12h 34m
Declination:	+7° 42′
Constellation:	Virgo
Object type:	Cataclysmic variable (Type Ia supernova)
Magnitude:	11.8 (at maximum)
Distance:	55 million light years (in galaxy NGC 4526)

Mira

Omicron (o) Ceti or Mira was the first 'true' variable star to be recognized, by Dutch astronomer David Fabricius in 1596. Its name means 'wonderful', and it is a red giant that pulsates between magnitudes 3 and 10 in a cycle of 332 days – its maximum and minimum brightness vary, but at its brightest it can reach magnitude 2.0. Mira and other pulsating stars are passing through a critical stage of evolution, where a slight change to their internal conditions turns a layer of their interior transparent or opaque. When the layer turns opaque, less light escapes from the interior, but the increased internal pressure makes the star billow outwards. As internal pressure and temperature drop, the internal layer turns transparent again, causing the pressure to drop and the star to collapse until the cycle repeats itself.

Formal designation:	o Ceti
Right ascension:	2h 19m
Declination:	-2° 59'
Constellation:	Cetus
Object type:	Variable star (pulsating red giant)
Magnitude:	3.04 (var)
Distance:	420 light years

RR Lyrae

This fast-changing variable star, best located with binoculars on Lyra's eastern edge, alters its brightness by a magnitude (between magnitudes 7.1 and 8.1) in a 13-hour period. It is the prototype for a class of variables known as RR Lyrae stars or cluster variables (because they are quite common in globular clusters). The star itself is a yellow-white giant about 50 times as bright as the Sun and 850 light years from Earth, passing through a period of instability as it nears the end of its life. RR Lyrae stars are important because, like Cepheid variables, the period of their oscillations is precisely linked to their luminosity, so it is easy to work out a star's distance by measuring its light curve and magnitude. RR Lyrae stars are too faint to see in other galaxies, but are a vital tool for mapping distances in the Milky Way.

Formal designation:	RR Lyrae
Right ascension:	19h 26m
Declination:	+42° 47′
Constellation:	Lyra
Object type:	Variable star (pulsating yellow-white giant)
Magnitude:	7.5 (var)
Distance:	855 light years

Delta Cephei

A variable star, Delta Cephei is a yellow supergiant in the northern constellation of Cepheus, roughly 900 light years from Earth. It ranges in brightness between magnitude 3.6 and magnitude 4.3 in a pattern that repeats every 5.37 days. The features of this 'light curve' are so distinct that they allow astronomers to identify other 'Cepheid' variables in the sky, even when they have different periods. This enabled US astronomer Henrietta Leavitt to prove, around 1912, that there is a precise relationship between the period of a Cepheid and its actual luminosity – so it is possible to work back from its light curve and its apparent magnitude to find its true distance from Earth. Since Cepheids are bright enough to be seen in other galaxies, this was an important first step in establishing the scale of the universe.

Formal designation:	δ Cephei
Right ascension:	22h 29m
Declination:	+58° 25′
Constellation:	Cepheus
Object type:	Variable star (pulsating yellow supergiant)
Magnitude:	4.7 (var)
Distance:	890 light years

R Coronae Borealis

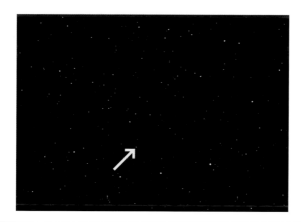

Most 'cataclysmic' variables, prone to sudden and dramatic changes, are novae that abruptly flare in brightness. A few stars, however, behave in the opposite way, and R CrB is certainly the most famous. It is a yellow supergiant that normally shines on the limit of naked-eye visibility at magnitude 5.9, but at unpredictable intervals it can plunge in brightness to around magnitude 15, taking weeks or months to recover. Analysis of the star's spectrum during these events reveals that its light is being blocked by clouds of carbon dust. The best theory to explain this behaviour is that the carbon is blown out of the star's upper atmosphere as vapour. At a certain distance from the star, it cools enough to condense into clouds, which block out the star's light until the pressure of its radiation forces them to disperse.

Formal designation:	R Coronae Borealis
Right ascension:	15h 49m
Declination:	+28° 09′
Constellation:	Corona Borealis
Object type:	Variable star (pulsating yellow supergiant with sudden fades)
Magnitude:	5.9 (var)
Distance:	6000 light years

Achernar

Achernar is the hottest of the sky's 'first-magnitude' stars (those with magnitudes greater than 1.5). Some 143 light years from Earth and six times the mass of the Sun, it is several thousand times more luminous. Its properties are complicated by a rapid rotation that gives it an equatorial bulge, as seen on gas- giant planets like Saturn. Across its equator, Achernar is 55 per cent wider than across its poles, meaning that the star is less dense around its equator. This makes the equatorial regions less dense and cooler than areas near the pole – a phenomenon known as gravity darkening. Temperatures on the star's surface range from 14,000°C (25,200°F) at the equator, to 19,000°C (34,200°F) at the poles. The entire star is wrapped in material flung off from its equator and shining in specific wavelengths.

Formal designation:	α Eridani
Right ascension:	1h 38m
Declination:	-57° 14'
Constellation:	Eridanus
Object type:	Rapidly spinning main- sequence blue star
Magnitude:	0.5
Distance:	143 light years

V838 Monocerotis

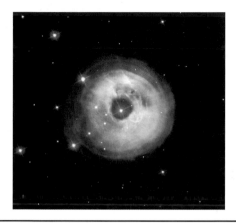

In 2002, a bizarre variable star burst into life: V838 leapt in brightness by several thousand times in a single day, fading and then brightening again twice more before returning to near its original brightness. Clouds of gas close to our sightline with the star then began to glow – so-called 'light echoes' caused as light reflected from these clouds reached Earth along its somewhat longer path. There are several theories to explain this spectacular outburst, but so far our lack of knowledge about the progenitor star and its environment has prevented a consensus being reached. Explanations range from the sudden ignition of a new phase of nuclear fusion inside a huge supergiant star, through a bizarre new nova, to a transformation caused by the violent merger of a pair of massive stars.

Formal designation:	V838 Monocerotis
Right ascension:	7h 04m
Declination:	-3° 51'
Constellation:	Monoceros
Object type:	Unusual cataclysmic variable supergiant star
Magnitude:	15.7–6.8
Distance:	About 20,000 light years?

Deneb

The brightest star in Cygnus is one of the sky's rare white supergiants – a monstrous star that shines at magnitude 1.2 in Earth's skies despite a distance of about 3000 light years. With a mass of about 20 Suns, Deneb is doomed to a short life of just a few million years: it has exhausted the supply hydrogen in its core, growing to a luminosity of 250,000 Suns in visible light alone as nuclear-fusion processes expand into a shell around the core. However, Deneb's powerful gravity counters the awesome pressure of this radiation, halting the expansion of its outer layers at around the diameter of Earth's orbit around the Sun. This ensures that the star's surface remains a white-hot 8000°C (14,500°F). In another few million years, Deneb will end its life in the blazing destruction of a supernova explosion.

Formal designation:	α Cygni
Right ascension:	20h 41m
Declination:	+45° 17'
Object type:	White supergiant star
Constellation:	Cygnus
Magnitude:	1.25
Distance:	3000 light years

Antares

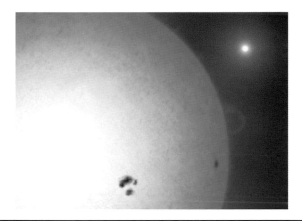

This bright red star in Scorpius (whose name means 'the rival of Mars') is a luminous supergiant star, 600 light years from Earth. With a mass of 16 Suns, it pumps out about 60,000 times the energy of the Sun in total, and the pressure of its internal radiation causes its outer layers to expand to a monstrous size. If Antares were placed at the centre of our solar system, it would engulf Earth and Mars, and stretch three-quarters of the way to the orbit of Jupiter. This means that its outer layers are extremely tenuous and cool (around 3300°C/5900°F), and as a result the star is vivid red and emits about 80 per cent of its energy as invisible infrared radiation. Strong stellar winds continually blow away gas from the outer atmosphere, forming a nebula that cocoons Antares and its blue companion star.

Formal designation:	α Scorpii
Right ascension:	16h 29m
Declination:	-26° 26'
Constellation:	Scorpius
Object type:	Variable red supergiant star
Magnitude:	1.0 (var)
Distance:	600 light years

The Pistol Star

Producing roughly 1.7 million times the energy of the Sun, the Pistol Star is one of the most luminous stars in the Milky Way. However, it is so distant – close to the centre of the galaxy and 25,000 light years from Earth – and hidden behind so much intervening gas and dust that it can be seen only in infrared light. The star is embedded in the heart of the Pistol Nebula in Sagittarius – an expanding cloud with about 10 times the mass of the Sun, containing material that was shrugged off by the star several thousand years ago (perhaps in a violent 'supernova impostor' event similar to the one that shook Eta Carinae in 1843). Born with the mass of 100 Suns, the Pistol Star is doomed to a short life of just a few million years, and its fierce radiation is blowing out a stellar wind with 10 billion times the strength of the Sun's.

Formal designation:	The Pistol Star
Right ascension:	17h 46m
Declination:	-28° 50'
Constellation:	Sagittarius
Object type:	Unusual cataclysmic variable supergiant star
Magnitude:	N/A (obscured)
Distance:	25,000 light years

Eta Carinae

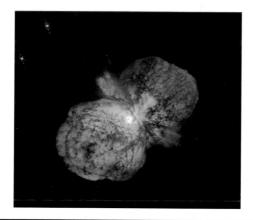

One of the most monstrous and violent stars in the sky, Eta (η) Carinae lies at the heart of the Carina Nebula. Beyond naked-eye brightness at magnitude 6.2, it is largely obscured by the twin lobes of the Homunculus Nebula surrounding it. This is an hourglass-shaped cloud of expanding material, flung out during an enormous explosion seen in 1843. At this time, the star brightened rapidly, to briefly outshine Canopus and become the second brightest star in the sky, before fading away. Eta Carinae is thought to be a monstrous star hurtling towards the end of its life and approaching the point where it will explode in a supernova. If Eta is a single star, then it must have a mass of 100 Suns, but there is also evidence that it might be a tightly bound pair of stellar monsters, each with a mass of 60–80 Suns.

Formal designation:	η Carinae
Right ascension:	10h 45m
Declination:	-59° 41'
Constellation:	Carina
Object type:	Binary blue supergiant star
Magnitude:	6.2 (var)
Distance:	8000 light years

Supernova 1987A

In 1987, a supernova explosion in the Large Magellanic Cloud* marked the death of a supergiant, Sanduleak -69° 202a. A Type II Supernova, it was the brightest, closest such event to occur since the invention of the telescope. Supernovae happen when a heavyweight star reaches the end of its life. Most stars dissolve into planetary nebulae as they exhaust their supply of helium, but high-mass stars keep shining by the fusion of carbon, nitrogen, oxygen and other elements. Each process creates heavier elements, but fusing heavier elements releases less energy, until the star tries to fuse iron, which absorbs energy rather than releasing it. The star's outer layers now collapse, compressing the core until a rebounding shockwave rips the star apart, igniting its outer layers in a burst of fusion that may take months to fade.

Formal designation:	SN 1987A
Right ascension:	5h 35m
Declination:	-69° 16′
Constellation:	Dorado
Object type:	Unusual Type II supernova
Magnitude:	3.0 (at maximum)
Distance:	168,000 light years

The Crab Nebula

The most prominent supernova remnant in the sky, the Crab Nebula is listed as M1 in Charles Messier's catalogue of potentially confusing, comet-like objects. It is the expanding cloud of gas created by a supernova explosion seen around the world in 1054. Recorded by astronomers from China to North America, it was briefly the brightest object in the sky apart from the Sun and Moon, shining four times as brightly as Venus. The patch of light that marks the site was first observed in 1754 by British astronomer John Bevis, and is 10 light years across. The weblike pattern of glowing gas filaments changes rapidly, altering the nebula's appearance across the decades. While the gas cloud is a source of X-ray and radio waves, the collapsed neutron star at its centre is a pulsar* – a source of pulsing radio waves.

Formal designation:	M1/NGC 1952
Right ascension:	5h 35m
Declination:	+22° 01'
Constellation:	Taurus
Object type:	Supernova remnant
Magnitude:	8.4
Distance:	6300 light years

Cassiopeia A

The strongest radio signals from the sky beyond our solar system come from a distant shell of glowing gas in Cassiopeia, about 10,000 light years away and 10 light years across. This is the glowing remnant of the most recent supernova explosion in our own galaxy, and is generally called Cassiopeia A. Disappointingly faint in visible light, the supernova remnant's light can be captured only by large telescopes, but is hot enough, at an astounding 30 million °C (54 million °F) to be a bright source of X-rays as well as radio waves. Measurements of the gas shell's rate of expansion suggest that the supernova exploded in around 1667, but there is no record of the event being observed, which suggests that the star was embedded in a thick cloud of opaque dust and gas thrown off during the later stages of its life.

Formal designation:	Cassiopeia A
Right ascension:	23h 23m
Declination:	+58° 48'
Constellation:	Cassiopeia
Object type:	Supernova remnant
Magnitude:	Unmeasured (extremely faint in visible light)
Distance:	10,000 light years

Cygnus Loop

One of the sky's brightest supernova remnants is the Veil Nebula in Cygnus, the most prominent part of a shell of gas called the Cygnus Loop. Extending over three degrees of the sky, this is a 'double' remnant, 2600 light years from Earth. The initial supernova exploded about 18,000 years ago, and as its gas blasted outwards, it hollowed out a spherical hole in the surrounding gas clouds, making them glow. Then, some 5000 years ago, a new supernova explosion within the remnant of the first created a second expanding layer of gas. Different colours in the loop trace the presence of various elements. While hydrogen, helium, carbon, oxygen and sulphur are common, small amounts of heavier elements can be created during the explosion itself – supernovae are the source of the universe's heaviest elements.

Formal designation:	NGC 6960, 6979, 6992, 6995
Right ascension:	20h 51m
Declination:	+30° 40'
Constellation:	Cygnus
Object type:	Supernova remnant
Magnitude:	5.0 (extremely diffuse)
Distance:	2600 light years

Vela Pulsar

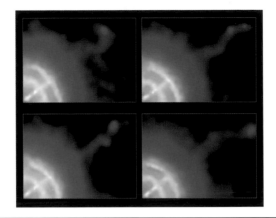

Supernova explosions put extreme pressure on a star's core, pushing its already densely packed particles further together. Usually, matter can be compressed only so far – pressure between electrons fights back, and a low-mass star's core stabilizes as a white dwarf. In a supernova, however, particles within the core are compressed to form a soup of neutrons that can be more tightly packed. The result is a neutron star – a city-sized ball of matter with the mass of several Suns. Neutron stars often have intense magnetic fields that concentrate radiation and high-energy particles into beams which emerge from the star's magnetic poles. If these are not aligned with the star's axis of rotation, the result is a pulsar – a star generating a pair of 'lighthouse beams' that seem to flicker on and off when they line up with Earth.

Formal designation:	PSR 0833-45
Right ascension:	8h 35m
Declination:	-45° 10′
Constellation:	Vela
Object type:	Supernova remnant/pulsar
Magnitude:	12 (for Vela supernova remnant)
Distance:	800 light years

Crab Pulsar

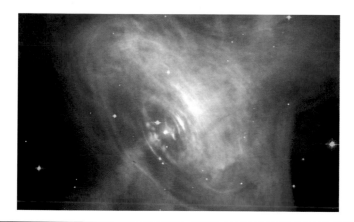

The Crab Pulsar was the first of its kind to be identified in visible light – previous pulsars were known only by radio signals. It flashes as its magnetic field rotates once every 33.1 milliseconds, and is gradually slowing as the star loses energy to its surroundings. Pulsars spin so rapidly because they retain the same rotational momentum as the original stellar core, in a much smaller object. Combined optical and X-ray images of the heart of the Crab Nebula* reveal the interaction between the pulsar at its heart and the surrounding supernova remnant. Strong stellar winds blowing off the neutron star's surface create rippled shock-fronts in the material near the star, while jets of particles blown out with the radiation from the poles and travelling close to the speed of light generate radiation at almost all wavelengths.

Formal designation:	PSR B0531+21
Right ascension:	5h 35m
Declination:	+22° 01′
Constellation:	Taurus
Object type:	Supernova remnant/pulsar
Magnitude:	16.0
Distance:	6300 light years

Black hole

R arely, the death of a massive star in a supernova explosion produces an object stranger than a neutron star. Just as the pressure between electrons gives way when a collapsing stellar core weighs more than 1.44 solar masses, so too the pressure between neutrons is not insurmountable. If a star's core weighs more than five solar masses (corresponding to an overall mass for the star of 20 or more Suns), the forces of its collapse shred its neutrons into their constituent particles, and the star collapses to a single point in space. Its gravity is so strong that nothing, not even light, can escape from it, and it is cut off from the universe within an impenetrable 'event horizon'. The star has become a black hole – an object that distorts space and time around it, and can be recognized only by the effect of its gravity.

Formal designation:	M33 X-7
Right ascension:	1h 34m
Declination:	+30° 32'
Constellation:	Triangulum
Object type:	70 solar-mass black hole in binary system
Magnitude:	N/A (detected through X-rays)
Distance:	3 million light years (in galaxy M33)

Cygnus X-1

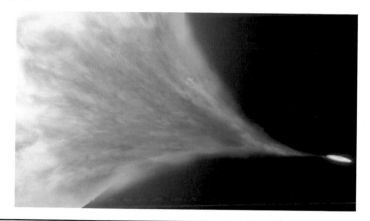

Black holes floating alone in space are almost impossible to detect – but they do sometimes become obvious in binary-star systems. Cygnus X-1 is the first potential black hole to be detected. It consists of a blue supergiant with the mass of 20 Suns, orbited once every 5.6 days by a supernova remnant of about five solar masses – all that remains of a more massive progenitor star. The remnant cannot be seen, but its mass is above the theoretical upper limit for a neutron star, so it must be a black hole, with gravity pulling away material from the supergiant's outer atmosphere. As gas spirals down into an 'accretion disc' around the hole, the increasing gravity creates tremendous tides that heat the disc. As the black hole shreds material around it, the disc emits X-rays that can be detected from Earth.

Formal designation:	Cygnus X-1
Right ascension:	19h 58m
Declination:	+35° 12′
Constellation:	Cygnus
Object type:	Black hole in binary system
Magnitude:	8.95
Distance:	8200 light years

Beta Pictoris

This white star, 63 light years from Earth, was the first star to be found with a disc of potentially planet-forming material around it. It was noticed because it emitted unusual amounts of infrared radiation; white stars, with surfaces hotter than the Sun, release most of their energy as visible or ultraviolet radiation, while infrared is produced by cooler objects. The first infrared images of the star showed it to be surrounded by a disc of cool material, causing its unusual behaviour. The disc has a diameter about 40 times that of Pluto's orbit around the Sun, which suggests that it could be a Kuiper Belt, similar to our own solar system's, in the act of formation around the still-young star. Recent images have shown gaps and distortions in the disc, perhaps created by young giant planets in orbit around the star.

Formal designation:	β Pictoris
Right ascension:	5h 47m
Declination:	-51° 04′
Constellation:	Pictor
Object type:	Star with extrasolar planetary system
Magnitude:	3.9
Distance:	63 light years

Fomalhaut

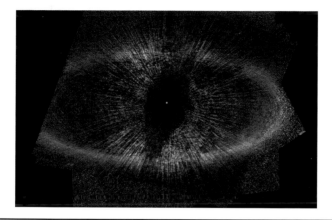

Several of the sky's brightest stars are surrounded by large, potentially planet-forming discs of gas and dust. Vega* is one, Fomalhaut in Piscis Austrinus is another. The star itself is 25 light years away, with a hot white surface, and emits about 16 times the light of the Sun. It is also fairly young, just 300 million years old. Fomalhaut's disk was discovered in the 1980s, but it took the power of the Hubble Space Telescope to reveal its structure in detail. The disc turned out to be a broad doughnut-shape, with a sharp inner edge 133 astronomical units from its star. The inner cutoff may indicate that planets have formed closer to the Sun in an area corresponding to the planetary region of our solar system, soaking up the material near the star. If so, the dusty disc is Fomalhaut's equivalent of our Kuiper Belt.

Formal designation:	α Piscis Austrini
Right ascension:	22h 58m
Declination:	-29° 37'
Constellation:	Piscis Austrinus
Object type:	Star with extrasolar planetary system
Magnitude:	1.2
Distance:	25 light years

51 Pegasi

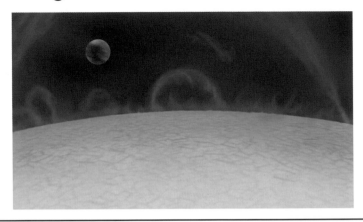

Just six per cent heavier than our own Sun and 30 per cent more luminous, 51 Pegasi lies 50 light years away, and is visible to the naked eye at magnitude 5.5. It hosts the first extrasolar planetary system discovered around a normal star. The planet, 51 Pegasi b, was announced in 1995 after astronomers measured the spectrum of 51 Pegasi and discovered a periodic shifting back and forth of the lines in its spectrum. This 'Doppler-shift' effect is caused by the star wobbling back and forth relative to Earth, as it is pulled this way and that by the gravity of a fairly heavy planet. Ultimatedly 51 Pegasis b proved to have a mass at least half that of Jupiter, and an orbital period of just 4.2 days at 0.05 astronomical units from its star – it is the prototype for a class of extrasolar planets known as 'hot Jupiters'.

Formal designation:	51 Pegasi
Right ascension:	22h 57m
Declination:	+20° 46'
Constellation:	Pegasus
Object type:	Star with extrasolar planetary system
Magnitude:	5.5
Distance:	50 light years

Tau Bootis

One of the most intensively studied extrasolar planets orbits the white star Tau (τ) Bootis, 51 light years from Earth. Tau itself is a double star, consisting of a white primary about 1.3 times the mass of the Sun, accompanied by a red dwarf in an elliptical orbit. The planet orbiting the primary star was discovered by its effect on the star's spectrum. It is a 'hot Jupiter' – a giant at least 3.9 times the mass of Jupiter, orbiting so close to its star that it completes a circuit in just 3.3 days. Using sensitive instruments, astronomers made the first direct measurement of light reflected off an extrasolar planet in 1999, revealing that it has a blue-green colour. It was recently discovered that planet and star are tidally locked, so each keeps one face permanently toward the other, like Pluto and Charon in our solar system.

Formal designation:	τ Bootis
Right ascension:	13h 47m
Declination:	+17° 27'
Constellation:	Bootes
Object type:	Star with extrasolar planetary system
Magnitude:	4.5
Distance:	51 light years

PSR B1257+12

Although most planets have been detected around 'normal' stars, the first extrasolar planetary system was found orbiting a long-dead neutron star. It was detected in 1992 when astronomers realized that the pulsar PSR B1257+12 was not keeping accurate time. Pulsars are rapidly spinning neutron stars that produce a regular signal as they rotate. This pulsar, however, was speeding up and slowing down in a repeating cycle, suggesting that something was pulling it around, and astronomers deduced the existence of three planets and a small 'comet' in orbit around the stellar remnant. Since astronomers have long assumed that no planet could survive the supernova explosion needed to create a neutron star, the 'pulsar planets' may have formed since the explosion, condensing out of a ring of debris.

Formal designation:	PSR 1257+12
Right ascension:	13h 00m
Declination:	+12° 40'
Constellation:	Virgo
Object type:	Pulsar with extrasolar planetary system
Magnitude:	N/A (detected from radio signals)
Distance:	980 light years

Gliese 581c

Discovered in 2007, Gliese 581c was the first truly Earthlike planet to be found, with conditions that have the potential to foster life. The planet is a 'super-Earth', a rocky (or possibly watery) planet with the mass of at least five Earths but (if it is indeed rocky) a similar diameter. It is one of three planets orbiting the red-dwarf star Gliese 581, just over 20 light years from Earth. If this star were placed in the middle of our solar system, the Earth would freeze, but Gliese 581c orbits close enough to its feeble star (in a distorted, Mercury-like orbit with an average radius of about 0.08 astronomical units) that its temperature is raised to an average of -3°C (27°F). Its atmosphere, however, could easily trap heat and help keep the planet warm enough for liquid water to exist on the surface.

Formal designation:	Gliese 581
Right ascension:	15h 19m
Declination:	-07° 43'
Constellation:	Libra
Object type:	Star with extrasolar planetary system
Magnitude:	10.6
Distance:	20 light years

2M1207

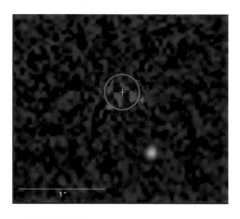

Extrasolar planets shine only through light reflected from their stars, and may be so dim in comparison that their images are lost in the glare. An exception is 2M1207b – the first planet found orbiting a brown-dwarf star, and probably the first extrasolar planet to be photographed. The brown dwarf itself, 2M1207, is thought to be very young, more than 170 light years away in the direction of Centaurus and about 20 times the mass of Jupiter. It was discovered by the 2MASS infra-red survey of the sky, and when astronomers at the European Southern Observatory in Chile observed it in 2004, they also detected light from a hot companion with about five times the mass of Jupiter. Astronomers still argue over 2M1207b's status – it may be a smaller brown dwarf in its own right, rather than a planet.

Formal designation:	2MASSW J1207334-393254
Right ascension:	12h 08m
Declination:	-39° 33'
Constellation:	Centaurus
Object type:	Brown dwarf with extrasolar planetary system
Magnitude:	13
Distance:	173 light years

OGLE-2005-BLG-390Lb

This planet, with an unfortunately long-winded name, is one of the furthest from Earth to be discovered, and the first success for a new planet-hunting technique. It is also noteworthy as the first rocky or icy extrasolar planet found – until 2006 all known planets around other stars were thought to be gas giants. 390Lb orbits a distant red dwarf 21,500 light years away, and was found through an effect called 'gravitational microlensing', caused when the gravity of a planet around a star deflects the path of light from another star. The planet has about 5.5 times the mass of the Earth, and could be either a truly rocky world like the Earth, or a smaller version of an ice giant such as Uranus. Unfortunately, it orbits between two and four astronomical units from its feeble star, and so is perpetually frozen.

Formal designation:	OGLE-2005-BLG-390L
Right ascension:	17h 54m
Declination:	-30° 23'
Constellation:	Sagittarius
Object type:	Red dwarf with extrasolar planetary system
Magnitude:	15.7
Distance:	21,500 light years

COROT-Exo-1b

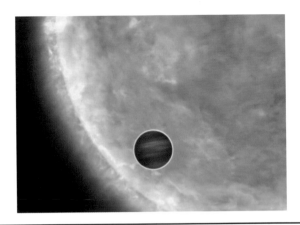

Discovered in 2007, this planet around a star in Monoceros, 1500 light years from Earth, was the first success for a French-led satellite aiming to use eclipses or planetary transits to detect extrasolar planets. COROT is equipped with sensitive photometers to measure tiny variations in the light of distant stars. It can detect a brightness dip equivalent to a planet a few times larger than Earth (potential 'super Earths') passing in front of its star, and also minute tremors in the surface of the stars themselves. COROT's transit-hunting method enables astronomers to measure the planet's size more or less directly. This showed that while COROT-Exo-1b is just 30 per cent more massive than Jupiter, its diameter is 80 per cent greater, since the hot conditions close to its sunlike star cause its atmosphere to expand.

Formal designation:	Corot-exo-1
Right ascension:	About 7h 00m
Declination:	About 0° 00′
Constellation:	Monoceros
Object type:	Star with extrasolar planetary system
Magnitude:	13.5
Distance:	1500 light years

The Milky Way

Our solar system and all of the stars normally visible to the naked eye are part of a single vast galaxy – a rotating star system about 100,000 light years across, in the shape of a flattened disc with a central bulge and spiral arms running across the disc. Because the solar system lies within the disc, when we look across the plane of the galaxy the light from distant stars blurs to form a broad band across the sky – the Milky Way that gives our galaxy its name. The star clouds are at their brightest and densest towards the constellation Sagittarius, which lies towards the galaxy's centre. The Sun lies between the Orion and Perseus spiral arms, roughly halfway between the heart of the Milky Way and its edge. It is just one of roughly 200 billion stars, and orbits our galaxy's central hub in about 240 million years.

Formal designation:	The Milky Way Galaxy
Position:	N/A
Object type:	Barred spiral galaxy
Mass:	580 billion solar masses
Number of stars:	About 200 billion
Diameter:	About 90,000 light years
Distance:	26,000 light years to centre

Sgr A*

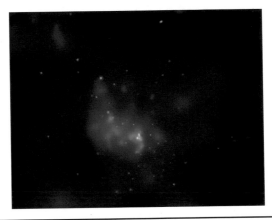

This radio source (pronounced 'Sagittarius A-star') is our only indication of the monstrous secret at the centre of the Milky Way galaxy. Astronomers believe it emanates from sparse gas clouds sifting down into the grasp of a 'supermassive' black hole, with the mass of perhaps four million Suns. Since no radiation can escape the black hole's grasp, it cannot be seen directly, even with infrared cameras that can look through the nearby clouds of stars, gas and dust and into the heart of the galaxy. It reveals itself by affecting the stars round it: their rapid orbits indicate a huge mass just a couple of astronomical units across. Astronomers now think that most galaxies contain supermassive black holes in their centre – they probably form from collapsing gas clouds early in each galaxy's evolution.

Formal designation:	Sgr A*
Right ascension:	17h 46m
Declination:	-29°
Constellation:	Sagittarius
Object type:	Supermassive black hole
Magnitude:	N/A
Distance:	26,000 light years

Omega Centauri

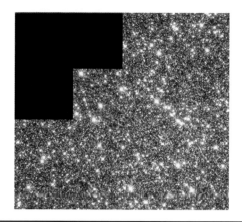

The finest globular cluster in the sky is NGC 5139 – a cluster so bright it has been granted a Greek letter of the type usually reserved for stars. Omega Centauri is one of the closest globulars to Earth, at a distance of 16,000 light years, and is also the largest cluster orbiting the Milky Way. It contains about 10 million stars, averaging half a solar mass each, crammed into a ball 150 light years across. Easily visible to the naked eye, it was first recorded by British astronomer Edmond Halley during an expedition to the southern hemisphere in 1677. The cluster's size (10 times more massive than the next largest Milky Way globular cluster) and some unusual properties of its stars lead some astronomers to think that Omega Centauri may actually be the remains of a galaxy consumed by the Milky Way.

Formal designation:	NGC 5139
Right ascension:	13h 27m
Declination:	-47° 29'
Constellation:	Centaurus
Object type:	Globular cluster
Magnitude:	3.7
Distance:	16,000 light years

47 Tucanae

One of the finest globular clusters, 47 Tucanae (also known as NGC 104) contains about 100,000 stars packed into a volume of space 120 light years across. Since it appears to the naked eye as a blurred 'star' of fourth magnitude, it bears a 'Flamsteed number' designation. Binoculars or a small telescope transform it into a densely packed ball of stars the size of the full Moon. Stars in globular clusters are usually red or yellow, with low masses and small amounts of heavier elements in their atmospheres. Similar to the stars in the hub of the Milky Way, they are often called Population II stars (but are thought to be older than the 'Population I' stars in the Milky Way's disc and spiral arms). Globular clusters are as old the Milky Way, and astronomers think that they all formed in a single wave of star formation.

Formal designation:	NGC 104
Right ascension:	0h 24m
Declination:	-72° 05'
Constellation:	Tucana
Object type:	Globular cluster
Magnitude:	4.9
Distance:	13,400 light years

M13

This globular cluster in Hercules is the brightest in northern skies, and one of the largest clusters orbiting the Milky Way: it contains perhaps a million stars in a ball 150 light years across. Just visible to the naked eye on a moonless night, it is best observed through binoculars or a small telescope, which reveal a densely packed core surrounded by ragged edges in which individual stars can be resolved. Although globulars like M13 are typically dominated by long-lived red and yellow stars, the presence of a handful of more luminous blue stars in their dense cores contradicts the theory that all globular stars formed in bursts billions of years ago. These 'blue stragglers' may be formed by the collision of two lower-mass red stars, which creates a star with higher mass and luminosity, and a hotter, bluer surface.

Formal designation:	M13/NGC 6205
Right ascension:	16h 42m
Declination:	+36° 28'
Constellation:	Hercules
Object type:	Globular cluster
Magnitude:	5.8
Distance:	25,100 light years

Sagittarius Dwarf Elliptical Galaxy

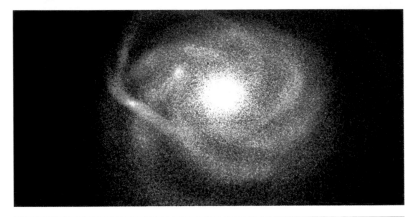

One of the closest galaxies to the Milky Way, the Sagittarius Dwarf Elliptical, or SagDEG, was discovered in 1994 when astronomers surveying star clouds above the centre of the Milky Way noticed an unusual number of low-mass red and yellow stars in a large region of the sky (5 x 10°). Statistics showed that this was unlikely to be random chance, and measurements of the distance to this star cloud showed that it lay 90,000 light years from Earth, above the far side of the Milky Way. It is, in fact, a small, highly distorted elliptical galaxy in the act of colliding with our own galaxy. SagDEG will be torn apart by its encounter, and its stars scattered through the galactic 'halo' above and below the hub of the Milky Way, but several globular clusters associated with it will probably survive more or less intact.

Formal designation:	SagDEG
Right ascension:	18h 55m
Declination:	-30° 33'
Constellation:	Sagittarius
Object type:	Dwarf elliptical galaxy
Magnitude:	4.5 (extremely diffuse)
Distance:	65,000 light years

Large Magellanic Cloud

Several irregular and elliptical galaxies orbit the Milky Way, but the largest and brightest is the Large Magellanic Cloud (LMC). Like the Small Magellanic Cloud, it was reported to European astronomers by Ferdinand Magellan, who noted it during his round-the-world voyage of 1619. Lying in the southern constellation of Dorado, it is easily visible to the naked eye, looking like a detached segment of the Milky Way. In reality, the LMC is a large irregular galaxy, some 30,000 light years across and 180,000 light years from Earth. Like many galaxies of its type, it is rich in clouds of gas and dust, and contains impressive areas of star formation, including the Tarantula Nebula. It has hints of large-scale structure, including a central bar of stars and what may be a single stunted spiral arm.

Formal designation:	Large Magellanic Cloud (LMC)
Right ascension:	5h 24m
Declination:	-69° 45′
Constellation:	Dorado/Mensa
Object type:	Irregular/peculiar spiral galaxy?
Magnitude:	0.1
Distance:	179,000 light years

Tarantula Nebula

Known by the designation 30 Doradus, the Tarantula is a star-forming region in the Large Magellanic Cloud*, visible to the naked eye as a bright patch amid its glowing starfields. Larger than any such region in the Milky Way, it is 1000 light years across, and so bright that if it lay at the same distance as the Orion Nebula M42, it would cast shadows at night. Binoculars reveal tendrils of gas in a spider-like shape, while a small telescope reveals star clusters embedded in the nebula, including R136a at its centre and Hodge 301 on its northwestern edge. These clusters contain some of the most massive stars known, and demonstrate how open clusters migrate as they age – Hodge 301 probably formed in the same region as the young R136a, but has drifted 150 light years in its 20 million years of life.

Formal designation:	NGC 2070
Right ascension:	5h 39m
Declination:	-69° 06′
Constellation:	Dorado
Object type:	Emission nebula
Magnitude:	8.0
Distance:	179,000 light years

Small Magellanic Cloud

A less impressive companion to the larger cloud in Dorado, the Small Magellanic Cloud (SMC) is another satellite of the Milky Way. Visible to the naked eye in the southern constellation of Tucana, it is fainter because it is smaller and further away (200,000 light years from Earth). It is a shapeless cloud of stars, gas and dust, rich in star-forming nebulae and young, brilliant blue stars. The Magellanic Clouds orbit the Milky Way in about a billion years, with the LMC leading and the SMC trailing. Our galaxy raises powerful tides in both, which may help to compress gas and trigger new waves of star formation, but it also takes its toll on its satellites, tearing away stars and interstellar material at each encounter, which spread out to form a faintly glowing trail, the Magellanic Stream, along the clouds' orbit.

Formal designation:	Small Magellanic Cloud (SMC)
Right ascension:	0h 53m
Declination:	-72° 50′
Constellation:	Tucana
Object type:	Irregular galaxy
Magnitude:	2.3
Distance:	210,000 light years

Andromeda Galaxy

This spiral of stars, gas and dust is the closest major galaxy to our own, and at a distance of three million light years, is the furthest object visible to the naked eye. Also known by its Messier catalogue label of M31, Andromeda is larger than the Milky Way (but probably not as massive), and dominates its own system of satellite galaxies, including the ellipticals M32* and M110 (NGC 205). Together, M31 and the Milky Way dominate an area of space 10 million light years across, creating a galaxy cluster called the Local Group. Viewed from Earth, M31's spiral arms blur into an ellipse of stars. Long-exposure photographs show dust lanes around and between the spiral arms, and distinctions in colour between the blue and white stars whose light dominates the disc, and the red and yellow stars in the bulbous central hub.

Formal designation:	M31/NGC 224
Right ascension:	00h 43m
Declination:	+41° 16'
Constellation:	Andromeda
Object type:	Spiral galaxy
Magnitude:	3.4
Distance:	2.5 million light years

Andromeda Galaxy hub region

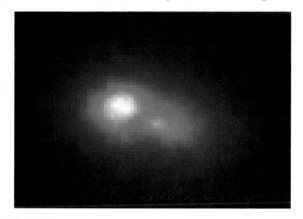

The core of Andromeda is the closest galactic hub that we can look at directly (even in our own Milky Way, the view is obscured by 20,000 light years of gas, dust and stars). Andromeda's central region is a flattened elllpse, 12,000 light years across and dominated by red and yellow stars that follow elliptical orbits tilted at various angles from the galaxy's horizontal plane. The orbits overlap to form an overall elliptical shape in the same way as happens for an elliptical galaxy. Infrared images of the hub suggest it may incorporate a bar of stars pointing directly away from us, and that M31, like the Milky Way, is a barred spiral. Image-processing techniques also reveal that the orbits of stars in the hub are systematically pulled 'off-centre' relative to the supermassive black hole around which they orbit.

Formal designation:	M31/NGC 224
Right ascension:	00h 43m
Declination:	+41° 16′
Constellation:	Andromeda
Object type:	Supermassive black hole in galactic hub
Magnitude:	N/A
Distance:	2.5 million light years

M32

Ellipticals are the most common type of galaxy in the universe, and include both the largest and the smallest galaxies. At one extreme lie the dwarf ellipticals, sparse balls of perhaps a few million well-scattered stars that can be very hard to detect. At the other lie monstrous giants like M87*, densely packed with trillions of stars. All lack star-forming dust and gas, and are dominated by long-lived red and yellow stars. The randomly tilted orbits of the stars combine to produce an overall elliptical structure. The two companions of M31 are among the brightest ellipticals in the sky, largely because they are so close to Earth. They are of an intermediate type – small, but still densely packed with several billion stars. M32 is the brighter and easier to find, since it lies directly in front of the larger spiral.

Formal designation:	M32/NGC 221
Right ascension:	0h 43m
Declination:	+40° 52′
Constellation:	Andromeda
Object type:	Elliptical galaxy
Magnitude:	8.1
Distance:	2.8 million light years

The Triangulum Galaxy

This face-on spiral galaxy, also known as M33, is the second closest major galaxy to the Milky Way – only the Andromeda Galaxy M31 is closer. Some three million light years from Earth, it is the third major member of our Local Group – but far smaller than our galaxy or M32. With a loose spiral structure, it falls into a group classed as 'flocculent' or 'clumpy'. Star formation is spread out into clouds throughout the galaxy's disc, rather than confined in sharply defined spiral arms like those seen in M101* and M51*. It seems that the spiral-forming density waves have weakened, allowing localized forces such as supernova shockwaves to control where and when new stars form. As a result, star formation spreads out in clumps, producing huge stellar construction sites such as the emission nebula NGC 604.

Formal designation:	M33/NGC 598
Right ascension:	1h 34m
Declination:	+30° 39′
Constellation:	Triangulum
Object type:	Spiral galaxy
Magnitude:	5.7
Distance:	3 million light years

Circinus Galaxy

This substantial spiral galaxy, 13 million light years from Earth, was found only in the 1970s, since it is largely hidden behind the Milky Way starclouds passing through the southern constellation of Circinus. It is the nearest example of an active galaxy – one whose overall energy output is greater than the sum of its stars. The Circinus Galaxy shows a sedate type of activity, mainly displayed in its brighter than expected core – it is therefore classed as a Seyfert galaxy (like M77*). However, images from the Hubble Space Telescope have revealed other features, such as a burst of starbirth in a broad ring of nebulae 700 light years from the centre (red in this false-colour image), and a lobe of hot gas (pink) apparently ejected from the galaxy's core, perhaps in the same way as the jets produced by radio galaxies.

Formal designation:	ESO 97-G13
Right ascension:	14h 13m
Declination:	-65° 20'
Constellation:	Circinus
Object type:	Seyfert galaxy
Magnitude:	12.1
Distance:	13 million light years

NGC 55

This large irregular galaxy is a borderline case in several ways. At seven million light years from Earth, it sits on the boundary between our Local Group of galaxies, and the neighbouring Sculptor Group centred on NGC 253. Once considered an outlying member of the Local Group, it is now included in the Sculptor Group: galaxy groups seldom have well-defined borders, and a galaxy may lie within one cluster yet 'belong' to another since its motion is determined by the second cluster's gravity. Although classed as irregular, NGC 55 is on the large side for one of these shapeless blobs of gas, dust and stars, and may be organizing itself: there are traces of a central bar of stars, a concentrated and bulbous 'hub', even the beginnings of spiral arms. Some astronomers describe the galaxy as spiral, not irregular.

Formal designation:	NGC 55
Right ascension:	0h 15m
Declination:	-39° 11'
Constellation:	Sculptor
Object type:	Spiral/irregular galaxy
Magnitude:	8.8
Distance:	7 million light years

Silver Coin Galaxy

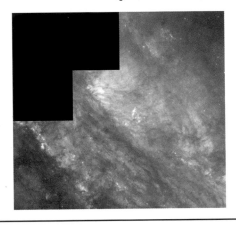

This beautiful spiral, catalogued as NGC 253 and sometimes known as the Sculptor Galaxy, is the dominant member of the Sculptor Group of galaxies, neighbours to our Local Group. It lies almost edge-on as seen from Earth, and is easily spotted through binoculars, although large telescopes are needed to reveal details of its structure. The galaxy has a relatively flat appearance, with spiral arms that are bright compared to its nucleus. Overall, it looks like a mottled oval of light, with darker patches indicating large dark dust clouds in the spiral arms. The Silver Coin is a 'starburst' galaxy undergoing a huge wave of star formation in the region around its core. The core itself is a source of X-rays and gamma rays, indicating that, like many other spirals, NGC 253 may be host to a central supermassive black hole.

Formal designation:	NGC 253
Right ascension:	0h 48m
Declination:	-25° 17
Constellation:	Sculptor
Object type:	Spiral galaxy
Magnitude:	7.1
Distance:	10 million light years

The Pinwheel Galaxy

The Pinwheel Galaxy was discovered by French astronomer Pierre Méchain in 1781 and became M101 in Charles Messier's catalogue. Almost twice the diameter of the Milky Way, at 170,000 light years across, this spiral has an uneven structure and a small, concentrated hub region at its centre. For such a large galaxy, it it is lightweight but dominates a large region of space because of its gravity, pulling half a dozen or more galaxies together to form the M101 Group. The central region can be spotted through binoculars, but large telescopes are required to see the spiral arms. Their asymmetric shape may be due to a collision with a companion galaxy in the relatively recent past, which may also have triggered waves of compression to create the many star-forming nebulae that dominate the spiral.

Formal designation:	M101/NGC 5457
Right ascension:	14h 03
Declination:	+54° 21'
Constellation:	Ursa Major
Object type:	Spiral galaxy
Magnitude:	7.9
Distance:	27 million light years.

NGC 1300

Roughly half of all spiral galaxies have an additional complexity to their structure in the form of a long straight bar of stars linking the central hub region to the spiral arms. In many galaxies (including the Milky Way), this 'bar' is simply an ovoid distortion of the hub. Others, such as NGC 1300, south of the celestial equator in the constellation Eridanus, have more spectacular bars. The bar of NGC 1300 extends across 80,000 light years – about half the diameter of the galaxy. Bars are thought to act as funnels that channel gas into the central hub and trigger bursts of star formation and other activity – the distortions may be caused by encounters with other galaxies. In NGC 1300's case, the system is more complex, since the centre contains its own small-scale spiral structure, about 3000 light years across.

Formal designation:	NGC 1300
Right ascension:	2h 20m
Declination:	-19° 25'
Constellation:	Eridanus
Object type:	Barred spiral galaxy
Magnitude:	11.4
Distance:	70 million light years

Sombrero Galaxy

This galaxy in Virgo, catalogued as M104, is an almost edge-on spiral with a large halo of stars and a dense dust ring around the outer edge of its spiral arms. Small telescopes show it as an oval blur of light; larger ones reveal the silhouetted dust lane; the most powerful reveal several hundred globular clusters orbiting the central hub. The Sombrero played a key role in our understanding of the scale of the universe. Vesto Slipher analyzed its spectrum of light in 1912 and found that various spectral lines were shifted to longer wavelengths than normal. Explaining this 'red shift' by the Doppler effect (in which light and sound waves lengthen when their source is receding from us), Slipher proved that M104 is moving away from us at 1000km (600 miles) per second – too fast to be an object within the Milky Way.

Formal designation:	M104/NGC 4594
Right ascension:	12h 40m
Declination:	-11° 37'
Constellation:	Virgo
Object type:	Spiral galaxy
Magnitude:	8.0
Distance:	50 million light years

NGC 4565

The closest spiral galaxy that is truly 'edge on' to the Milky Way, NGC 4565 lies in the Coma Berenices. At a distance of 30 million light years, it is closer than the Virgo or Coma galaxy clusters that spread across this constellation. Medium-sized telescopes will show that the ovoid blur of light from this galaxy's hub is bisected by a thick dust lane, and long exposure photographs will show the faint glow of light emerging from the galaxy's disc, some 100,000 light years across. Edge-on spirals reveal the true scale and shape of most spiral galaxies – the thin plane of the galaxy may be just a few hundred light years across, while the central hub may extend to several thousand light years, but is generally flatter than it is wide. They also reveal the true extend of dark dust clouds in the disc region of the galaxy.

Formal designation:	NGC 4565
Right ascension:	12h 36m
Declination:	+25° 59'
Constellation:	Coma Berenices
Object type:	Spiral galaxy
Magnitude:	9.6
Distance:	31 million light years

M60

Messier 60 in Virgo appears through telescopes as a fuzzy blob – it is simply an elliptical ball of stars, a bigger version of the ellipticals orbiting the Andromeda Galaxy M31*. Ellipticals generally contain little, if any, gas and dust to form new generations of stars. As a result, they are dominated by low-mass red and yellow stars, which are the only ones long-lived enough to have survived the billions of years since their formation. Each star follows its own elliptical orbit around the galaxy's centre of mass (typically marked by a supermassive black hole), and their overlapping orbits form an elongated elliptical ball. Ellipticals are classified as types from E0 (perfectly spherical), up to E7 (extremely elongated). Types E2–E3 are the most common, while the largest ellipticals are usually type E0.

Formal designation:	M60/NGC 4649
Right ascension:	12h 44m
Declination:	+11° 33'
Constellation:	Virgo
Object type:	Elliptical galaxy
Magnitude:	8.8
Distance:	60 million light years

NGC 2787

At first, this galaxy in Ursa Major seems elliptical; its dominant feature is an ovoid ball of old, red and yellow stars, and there is no sign of spiral arms extending into nearby space. But detailed photographs reveal the truth: although it has no spiral arms, NGC 2787 does have a disc of gas and dust similar to that in a spiral galaxy, populated with low-mass, faint stars. These same stars orbit in the disc of a normal spiral – indeed, the Sun is one of them – but are so faint that the disc seems empty between the bright spiral arms. NGC 2787, then, is a spiral without its arms, a lenticular (lens-shaped) galaxy. Astronomers think that many galaxies pass through a lenticular phase, perhaps as they recover from a galactic merger. Eventually their spiral arms can be reinvigorated, perhaps by an encounter with another galaxy.

Formal designation:	NGC 2787
Right ascension:	9h 19m
Declination:	69° 12'
Constellation:	Ursa Major
Object type:	Lenticular galaxy
Magnitude:	11.8
Distance:	24 million light years

Bode's Galaxy

This face-on spiral galaxy (one of two in Ursa Major), found by Johann Elert Bode in 1774, is a 'grand design' spiral, in which the spiral pattern is sharply defined. The nucleus is visible through binoculars, and reveals its ovoid shape to a small telescope, but larger instruments are needed to trace the arms. The spiral structure of certain galaxies is a mystery – it cannot be permanent, since a galaxy's centre always spins faster than its edge, and physical spiral arms would quickly 'wind up'. It seems the spiral is constantly renewed by the compression of interstellar gas and dust in spiral 'density waves' – galactic traffic jams that arise due to the overlap of countless individual orbits. The compressed regions create star-forming nebulae, which generate the open clusters of short-lived stars that mark out the visible spiral.

Formal designation:	M81/NGC 3031
Right ascension:	09h 56m
Declination:	+69° 04'
Constellation:	Ursa Major
Object type:	Spiral galaxy
Magnitude:	6.9
Distance:	12 million light years

The Cigar Galaxy

This irregular galaxy is close to M81 in Earth's skies, at 2–300,000 light years away. Indeed, its gravity may influence the structure of the larger spiral. The Cigar itself, M82 in Charles Messier's catalogue, is a typical 'starburst galaxy' – an irregular experiencing such a violent wave of star formation that it was once thought to be exploding. That may not be far from the truth – X-ray images and images taken by Hubble Space Telescope show plumes of hot gas blowing out of the galaxy's central region roughly perpendicular to its plane. It seems that M82 is experiencing runaway starbirth: huge stars forming rapidly in the centre (perhaps triggered by the gravity of M81) are joining forces to create a 'galactic wind' that ploughs the cold gas and dust further out in the galaxy, triggering more starbirth.

Formal designation:	M82/NGC 3034
Right ascension:	09h 56m
Declination:	+69° 41′
Constellation:	Ursa Major
Object type:	Irregular starburst galaxy
Magnitude:	8.4
Distance:	12 million light years

Malin-1

Discovered by accident in 1987, Malin-1 is the largest spiral galaxy yet found – a sparse, almost starless disk with slight hints of spiral structure. At 650,000 light years across, it dwarfs other spirals like the Milky Way and Andromeda. Ironically, its claim to fame is its near-invisibility: Malin-1 was the first large galaxy to be found with a low surface brightness. In other words, it produces the same light as a large galaxy, but this is spread out over a greater area, and so is difficult to detect. Despite this, the galaxy is a heavyweight: it contains 50 billion solar masses of gas and dust, far more than the Milky Way. In 2007, astronomer Aaron Barth discovered a 'normal' barred spiral at the centre of Malin-1, missed in earlier images because the entire galaxy lies more than a billion light years from Earth.

Formal designation:	Malin-1
Right ascension:	12h 37m
Declination:	+14° 20′
Constellation:	Coma Berenices
Object type:	Low surface brightness spiral galaxy
Magnitude:	25.7
Distance:	1 billion light years

Hoag's Object

One of the most beautiful but enigmatic galaxies in the heavens, Hoag's Object was found in 1950. At a distance of 600 million light years, it lies beyond the range of most amateur telescopes, but this image from the Hubble Space Telescope shows it has a disc-shaped core of old red stars, and an outer ring of younger stars, gas and dust. The space in between is not just dark, but empty, since more distant galaxies can be seen through it. Astronomers are still puzzling over the origins of Hoag's Object – it does not seem to be a true 'ring galaxy' like the Cartwheel*, or an extreme kind of spiral, as some have suggested. Perhaps the best theory is that it represents a galactic merger, a spiral or irregular galaxy that has been shredded and sent into a circular orbit around an older elliptical nucleus.

Formal designation:	PGC 54559
Right ascension:	15h 17m
Declination:	21° 35'
Constellation:	Serpens
Object type:	Peculiar galaxy
Magnitude:	16.0
Distance:	600 million light years

The Whirlpool Galaxy

This face-on spiral galaxy in Canes Venatici, catalogued by Charles Messier as M51, is particularly fine. Even across almost 40 million light years, it is bright enough for binoculars to identify its ovoid central hub. Details of the sharply defined spiral arms are revealed by a medium-sized telescope, as is the hub of a smaller irregular galaxy close by – NGC 5195. The two galaxies seem to be linked by a long stream of connecting material, but this is, in fact, an extension of one of the spiral arms, passing in front of the smaller companion. Once again, the strong spiral pattern of a 'grand design' spiral seems to be linked to an encounter with another galaxy. Astronomers believe that such interactions strengthen and invigorate the spiral density waves that trigger star cluster formation in the galaxy's arms.

Formal designation:	M51/NGC 5194
Right ascension:	13h 30m
Declination:	+47° 12′
Constellation:	Canes Venatici
Object type:	Spiral galaxy
Magnitude:	8.4
Distance:	37 million light years

M87

M87 is one of the largest galaxies known – a giant elliptical. It is a spherical ball of perhaps one million million stars at the heart of the Virgo Cluster of galaxies, with a diameter of half a million light years. Monsters such as M87 are thought to form from the repeated collision and merging of smaller galaxies in the centre of large galaxy clusters. Generally they contain little gas or dust and so have little new starbirth going on within, but they act as 'ballast' for a cluster, and hot gas stripped away during the merger process tends to fall towards this centre of gravity. M87 is unusual in that it has cannibalized another galaxy in the relatively recent past. The infall of gas and dust has fired M87's central black hole into life, producing a long jet of high-energy particles and turning the galaxy into a powerful radio source.

Formal designation:	M87/NGC 4486
Right ascension:	12h 31m
Declination:	+12° 24'
Constellation:	Virgo
Object type:	Giant elliptical radio galaxy
Magnitude:	8.6
Distance:	60 million light years

NGC 1316

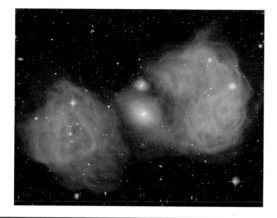

Coinciding with a strong radio source called Fornax A, there is a mottled lane of dust across the face of the lenticular galaxy NGC 1316, which bears no obvious relationship to the rest of the galaxy's structure. This is probably a relic of a dusty spiral that has largely been absorbed into the galaxy, transforming the black hole at its centre into a radio galaxy, with a pair of lobes extending to either side of the visible galaxy and covering several degrees of sky. According to the most popular theory of galaxy formation, lenticulars like NGC 1316 are, in fact, ellipticals (perhaps formed by mergers themselves) and are 'refueling' by pulling in gas and dust from the intergalactic medium around them. Eventually, if conditions are right, large-scale star formation may begin in the disc and spiral arms may even re-form.

Formal designation:	NGC 1316
Right ascension:	3h 23m
Declination:	-37° 13'
Constellation:	Fornax
Object type:	Lenticular radio galaxy
Magnitude:	9.4
Distance:	70 million light years

M77

This face-on spiral galaxy in Cetus is one of the finest in the sky, shining at magnitude 9.6 despite a distance of 60 million light years. A small telescope reveals that much of its brightness comes from a starlike point of light embedded in the galaxy's central hub. In fact, M77 is an example of a Seyfert galaxy – the least violent type of active galaxy, in which much of its radiation comes from its active core rather than its stars. Seyferts like M77 are believed to have supermassive black holes in their cores, which are feeding steadily on gas from their surroundings. As the gas is pulled in towards the hole, it forms a spiral 'accretion disc' that is heated up by friction and the enormous tides created by the black hole. The central light source of these galaxies is actually the surface of this accretion disc.

Formal designation:	M77/NGC 1068
Right ascension:	2h 43m
Declination:	0° 0'
Constellation:	Cetus
Object type:	Spiral Seyfert galaxy
Magnitude:	9.6
Distance:	60 million light years

Centaurus A

NGC 5128 is an unusual galaxy in Centaurus, visible through binoculars on a dark night. Telescope images reveal it as a type of elliptical galaxy – a huge ball of stars crossed by a dark lane of light-absorbing dust. The discovery of a strong radio source emanating from this same point in the sky led astronomers to catalogue it as Centaurus A. Today, astronomers think that NGC 5128 is an active galaxy, with a supermassive black hole at its centre, which is feeding on gas, dust and stars drawn in from the space around it, and spitting vast particle jets out of the galaxy. Where these encounter cooler gas in the space around the galaxy, they cause it to emit radio waves in two huge lobes. The dark trail of dust across the face of Centaurus A suggests that its activity has been triggered by consuming a smaller spiral galaxy.

Formal designation:	NGC 5128
Right ascension:	13h 26m
Declination:	-43° 01′
Constellation:	Centaurus
Object type:	Elliptical radio galaxy
Magnitude:	7.0
Distance:	15 million light years

Cygnus A

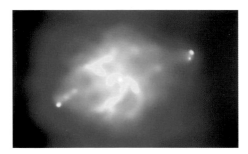

For radio astronomers, an area of the sky in Cygnus is dominated by two lobes of glowing gas, 120,000 light years across and 600 million light years from Earth. At the heart of this radio source, called Cygnus A, is a faint distorted galaxy, visible through large optical telescopes and connected to the radio lobes by a narrow jet of matter. The central galaxy is a single giant elliptical crossed by dust from a recently absorbed spiral, which has stirred its central black hole into life. Particles stream out in jets from around the black hole and plough into hot gas falling towards the galaxy because of its gravity. This creates a cavity that encloses the galaxy, with X-ray and radio-emitting 'bright spots' at either end where the particle jets spread out into turbulent lobes.

Formal designation:	3C 405
Right ascension:	19h 59m
Declination:	+40° 44'
Constellation:	Cygnus
Object type:	Elliptical radio galaxy
Magnitude:	15.0
Distance:	600 million light years

The Antennae

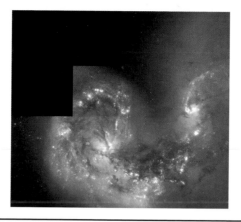

One of the sky's best-known pairs of colliding galaxies is located in Corvus, close to its border with Crater. The Antennae appear through small telescopes as smudges of light, but long exposures show a pair of stellar streamers extending from them, giving rise to their name. They seem to be spiral arms that have unwrapped themselves as two galaxies, NGC 4038 and 4039, approached each other. Images taken by the Hubble Space Telescope show the violent results of collision between two galaxy cores – huge clouds of gas and dust have slammed into each other and compressed to create a wave of new star formation, with at least 1000 massive new open clusters illuminating the central regions. Despite this violence, collisions between individual stars, which are relatively well spaced out, will be rare.

Formal designation:	NGC 4038/9
Right ascension:	12h 02m
Declination:	-18° 52'
Constellation:	Corvus
Object type:	Interacting spiral galaxies
Magnitude:	10.3
Distance:	63 million light years

The Mice

Named because their unwinding spiral arms give them a striking resemblance to a pair of white mice in the sky, this pair of galaxies is classified as NGC 4676. Like the Antennae*, they are caught in the act of collision and merger, but only the most powerful telescopes can reveal them in their full glory. Although collisions between individual stars during a galaxy merger are rare, clouds of gas and dust are forced into each other at high speeds. This triggers an immense burst of star formation, but also heats the gas to such high temperatures that it can escape the galaxies' gravity and 'boil away' into surrounding space. As the surviving stars from the merger settle into new and chaotic orbits, there is no material left to form new generations of stars. This is one theory for how large elliptical galaxies form.

Formal designation:	NGC 4676
Right ascension:	12h 46m
Declination:	-30° 44'
Constellation:	Coma Berenices
Object type:	Interacting spiral galaxies
Magnitude:	14.7
Distance:	300 million light years

The Cartwheel

This bizarre galaxy, 500 million light years away in Sculptor, was photographed by the Hubble Space Telescope in 1995. It revealed the tale of an apparent galactic hit and run, in which a smaller galaxy (one of its two close neighbours) has smashed through the centre of a large spiral and disrupted the spiral arm pattern, while also triggering tides that have caused a ring of star-forming activity to ripple out across the galaxy's disc, where it is now nearing the outer edge and forming a bright 'rim' some 150,000 light years across. The Cartwheel is the prototype for a class of 'ring galaxies' of similar origins. Since they lack spiral arms, they are classified as lenticulars. However, there are hints amid the gas, dust and fainter stars between the Cartwheel's rim and hub, that a spiral may be regenerating itself.

Formal designation:	ESO 350-G40
Right ascension:	0h 37m
Declination:	-33°43'
Constellation:	Sculptor
Object type:	Interacting spiral and irregular galaxies
Magnitude:	19.3
Distance:	500 million light years

3C 273

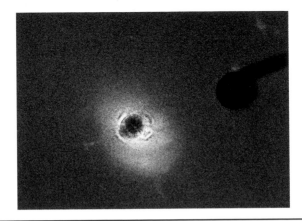

In the 1950s, astronomers began to find a new type of 'radio star' – starlike points of light that varied in brightness, emitted radio waves as well as visible light, and had mistifying spectra. 3C 273 in the constellation of Virgo was the brightest. In 1962 astronomers obtained a more detailed version of its spectrum, and realized its bright lines were similar to those in normal galaxies, but 'redshifted' by a vast amont, suggesting that the source was moving away from Earth at 47,000km (29,200 miles) per second. As a result, the radio stars were transformed into 'quasi-stellar objects', or quasars – violent sources of radiation billions of light years away. These starlike energy sources are often associated with distant galaxies, and quasars are a powerful form of active galaxy, similar to radio galaxies and Seyferts.

Formal designation:	3C 273
Right ascension:	12h 29m
Declination:	+02° 03'
Constellation:	Virgo
Object type:	Quasar in elliptical galaxy
Magnitude:	12.9
Distance:	2.4 billion light years

BL Lacertae

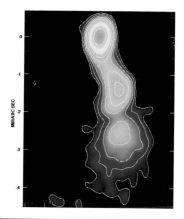

As with the first quasars, the distant galaxy BL Lacertae was at first mistaken for a variable star with a bizarre spectrum. It was only in 1968 that astronomers realized it was also a radio source and therefore probably an active galaxy. BL Lacertae soon became the prototype for a class of galaxies known as blazars or BL Lac objects. These galaxies show rapid variations in brightness, and almost featureless spectra. It is now thought that BL Lac objects are essentially similar to quasars and radio galaxies, and that all these galaxies – as well as the less powerful Seyferts – are powered by the same central 'engine' mechanism. The huge distances of many active galaxies mean that we see them as they were billions of years ago; they may, in fact, be a phase that all galaxies pass through in their youth.

Formal designation:	BL Lac
Right ascension:	22h 02m
Declination:	+42° 17′
Constellation:	Lacerta
Object type:	Blazar in elliptical galaxy
Magnitude:	15 (var)
Distance:	1 billion light years

NGC 7052

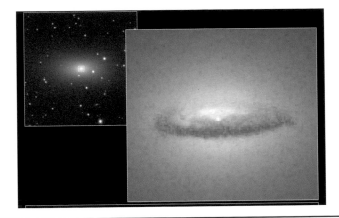

The 'standard model' of active galactic nuclei assumes that active galaxies have a similar central engine, revealed in this image taken by Hubble Space Telescope of the heart of radio galaxy NGC 7052. Seyferts, quasars, blazars and radio galaxies are driven by the energy from a black hole pulling gas, dust, and even stars to their doom. The material spiralling into the black hole forms an 'accretion disc' that is intensely heated to produce radiation, from visible light to X-rays. Meanwhile jets of particles are flung off at right-angles to the disc at close to the speed of light. A cloud of dust and gas obscures the disc from 'edge-on', and for radio galaxies we see only the effects of the jet billowing into space. In quasars and Seyferts we see radiation from the disc itself, while in blazars we are looking straight down the particle jet.

Formal designation:	NGC 7052
Right ascension:	21h 19m
Declination:	+26° 27′
Constellation:	Vulpecula
Object type:	Elliptical radio galaxy
Magnitude:	13.4
Distance:	190 million light years

Stefan's Quintet

This cluster of interacting galaxies in Pegasus shows the fate of galaxy groups like our own, and demonstrates the scale of the universe. The Quintet is actually a quartet of four galaxies some 300 million light years away, while the larger galaxy (NGC 7020) is in the foreground, at 40 million light years from Earth. The evidence for the different distances comes from redshift: since the universe is expanding, more distant objects retreat from us faster and therefore have greater redshifts in their light. The four interacting galaxies are the large members of a cluster similar to the Local Group, pulled towards each other by their own gravity. They will eventually merge, but as they disrupt each other's structures and raise tides of star formation, they have produced at least 100 star clusters and several dwarf galaxies.

Formal designation:	Hickson 92
Right ascension:	22h 36m
Declination:	+33° 58'
Constellation:	Pegasus
Object type:	Compact galaxy group
Magnitude:	13.9 and fainter
Distance:	300 million light years

Virgo Cluster

The closest major galaxy cluster to Earth lies about 60 million light years away in northern Virgo and southern Coma Berenices. It contains about 1000 galaxies, but is surprisingly compact – at 10 million light years across, it is barely larger than the Local Group. In fact, all galaxy clusters are about the same size. Around the edges of the Virgo Cluster lie large mixed spiral and elliptical galaxies, but its centre is dominated by a few giant ellipticals – huge balls of stars, including M87, the biggest of all. Compared to more distant clusters such as the Norma Cluster, Virgo's galaxies have a random distribution. Over billions of years, gravity evens out the distribution of galaxies in a cluster until it is roughly spherical or 'relaxed', so Virgo's chaotic spread suggests that it came together quite recently.

Formal designation:	Virgo I Cluster
Right ascension:	12h 30m
Declination:	+13°
Constellation:	Virgo/Coma Berenices
Object type:	Galaxy cluster (about 2000 galaxies)
Magnitude:	9.0 and fainter
Distance:	About 60 million light years

Norma Cluster

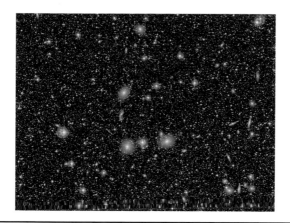

Partially obscured by the Milky Way, the Norma Cluster is an ancient and relaxed galaxy cluster. It lies near the centre of the Centaurus Supercluster, a 'cluster of clusters', about 250 million light years away. The Centaurus Supercluster seems to be associated with the object or region of space known as the Great Attractor – a concentration of mass that exerts its gravity even this far away, pulling the Virgo Cluster, and ourselves too, towards it at 2.2 million km (1.3 millions miles) per hour. The visible galaxies in this part of space cannot account for the gravity, suggesting that there is also a vast clump of invisible 'dark matter' in this part of space. Dark matter is the unseen bulk of the universe, accounting for up to 90 per cent of all matter. It reveals itself only by its gravitational effects, and remains mysterious.

Formal designation:	Abell 3627
Right ascension:	16h 15m
Declination:	-60° 55'
Constellation:	Norma
Object type:	Galaxy cluster (several thousand galaxies)
Magnitude:	N/A (obscured)
Distance:	About 250 million light years

Abell 2218

This distant and dense galaxy cluster demonstrates a fundamental concept of cosmology – Einstein's Theory of General Relativity. According to Einstein, the presence of matter bends space and time around it, which means that light passing close to a massive object is deflected as it passes through the 'gravitational well' around it. This phenomenon, known as 'gravitational lensing', is explained only by relativity; because light has no mass, gravity could not affect it in any other way. The bluish arcs of light around Abell 2218 show gravitational lensing in action: they are distorted images of a more distant galaxy cluster directly behind it. The lensing effect is valuable for observers, since it can magnify and intensify light, bringing galaxies that would otherwise be undetectable within range of powerful telescopes.

Formal designation:	Abell 2218
Right ascension:	16h 36m
Declination:	+66° 13′
Constellation:	Draco
Object type:	Galaxy cluster and gravitational lens
Magnitude:	17.0 and fainter
Distance:	About 2 billion light years

Hubble Deep Field

In 1995, astronomers turned the gaze of the Hubble Telescope on to an 'empty' area of Ursa Major for 10 days, creating the deepest view of the universe yet seen. The Hubble Deep Field revealed a universe packed with galaxies, up to 10 billion light years from Earth. The time taken for light to cross this gulf of space means that these distant galaxies appear as they were when the universe was less than four billion years old. Astronomers noted several features, including the 'blue excess': the more distant a galaxy is, the more its light should be redshifted, but when this is taken into account, there appear to be far more blue, gas and dust-rich galaxies than seen in today's universe. This suggests that galaxies show a trend from blue and irregular towards yellow and elliptical throughout the history of the universe.

Formal name:	The Hubble Deep Field (HDF)
Right ascension:	12h 37m
Declination:	+62° 13′
Constellation:	Ursa Major
Object type:	Field of galaxies
Magnitude:	N/A
Distance:	Up to 10,500 million light years

Hubble Ultra-Deep Field

Spurred by the success of the Hubble Deep Field, NASA scientists created an image with the combined light of 15.8 days of exposure. This time, they focused on a region of Fornax one-tenth the diameter of the full Moon, and found 10,000 galaxies, including the most distant yet seen. The most ancient are so far away that their light began its journey to us only a few hundred million years after the Big Bang itself, and include shapeless irregular galaxies and spirals forming from the mergers of irregulars. Among these, Hubble found 'monster' galaxies that seem to have grown surprisingly quickly, and are large even by the standards of today's galaxies. Rather than growing in a series of slow mergers, they probably formed in a single intense burst of star formation from a huge collapsing cloud of matter.

Formal name:	The Hubble Ultra Deep Field (HUDF)
Right ascension:	3h 33m
Declination:	-27° 47′
Constellation:	Fornax
Object type:	Field of galaxies
Magnitude:	N/A
Distance:	Up to 13,000 million light years

The large-scale universe

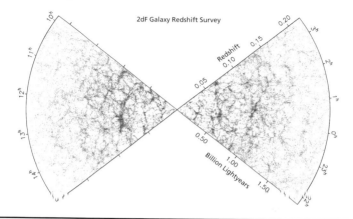

2dF Galaxy Redshift Survey

This diagram, plotting the positions of a quarter of a million galaxies in terms of their location in the sky and distance from the Milky Way (at the centre of the fan) gives us our clearest picture of the structure of the universe on the largest scales. Galaxies gather in clusters, which themselves form superclusters. (Our own Local Group is part of the Virgo Supercluster.) Superclusters run along chains known as filaments, with vast and apparently empty spaces between them called voids. Despite this structure, on the very largest scales of all, the universe looks roughly the same in all directions. Such maps bring home an important point: our location in space may be nothing special, but it is the centre of our 'observable universe', beyond which light has not had enough time to reach us since the Big Bang itself.

Obtained by:	2dF Galaxy Redshift Survey
Location:	N/A
Image type:	Plot of galaxy positions
Extent:	1500 square degrees
No. of galaxies measured:	232,155
Magnitude:	N/A
Distance:	Up to 1.96 billion light years

Cosmic Microwave Background

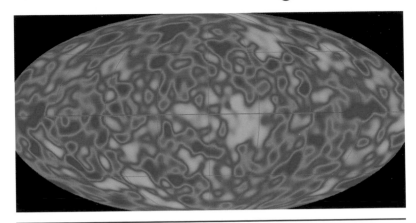

The edge of the observable universe - roughly 13.6 billion light years away in every direction, is too faint to see through a telescope. However, in 1963 two engineers using an ultra-sensitive horn radio antenna detected an unexpected glow of short radio waves (microwaves) coming from all over the sky. This proved to be the afterglow of creation itself – radiation from the moment the rapidly growing, incandescent ball of that was the infant universe became transparent for the first time. Discovery of the so-called 'Cosmic Microwave Background', now so ancient and feeble that it only warms the sky to 2.7° above absolute zero, clinched the case for the Big Bang theory that the universe originated in an enormous explosion.

Obtained by:	Cosmic Background Explorer
Location:	N/A
Image type:	Map of microwave radiation from 300,000 years after Big Bang
Average temperature:	2.725K (2.725°C above absolute zero)
Sensitivity:	1 part in 100,000
Magnitude:	N/A
Distance:	About 13.6 billion light years

312

The edge of the universe

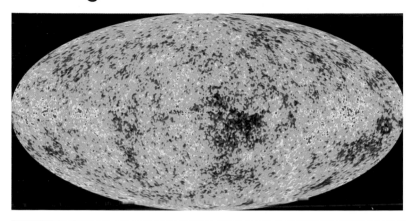

In 2001, NASA launched the Wilkinson Microwave Anisotropy Probe (WMAP), a sophisticated satellite designed to follow up on COBE's studies of the cosmic microwave background. The resulting map of the sky shows variations in the radiation's temperature of just millionths of a degree. These 'ripples' originate from irregularities in the infant universe: redder patches are hotter, indicating a greater density of material, while blue patches are cooler and less dense. The variations show that matter had already begun to 'clump' together while the universe was still opaque (it became transparent 300,000 years after the Big Bang). As the clumps and filaments continued to collapse, they eventually formed galaxy clusters and superclusters, and filaments surrounding the empty voids of the modern universe.

Obtained by:	Wilkinson Microwave Anisotropy Probe
Location:	N/A
Image type:	Map of microwave radiation from 300,000 years after the Big Bang
Average temperature:	2.725K (2.725°C above absolute zero)
Sensitivity:	1 part in 4.5 million
Magnitude:	N/A
Distance:	About 13.6 billion light years

Glossary

Accretion disc
A spiral disc of material formed around an object as it pulls matter from, for example, a gas cloud or companion star onto its surface.

Big Bang
The event in which the Universe originated as a tiny, unimaginably hot and dense ball of matter, roughly 13.6 billion years ago.

Binary star
A star system consisting of two stars in orbit around their common centre of mass (which is closer to, and sometimes inside, the more massive star).

Black hole
An object so dense that not even light can escape from its surface, and space and time are warped in its surroundings.

Brown dwarf
A 'failed star' with a mass of more than 12 Jupiters, which shines dimly through low-energy nuclear fusion of deuterium.

Dark matter
Unseen matter in the Universe that does not emit light or any other radiation, and makes its presence felt only through the influence of its gravity.

Dark nebula
A nebula that does not shine by emission or reflection, and can only be seen where it is silhouetted against a brighter background.

Dwarf planet
An object in an independent orbit around the Sun (or another star) that has sufficient mass to pull itself into a spherical shape, but whose gravitational influence is not strong enough to clear its orbit of other objects.

Emission nebula
A nebula that shines thanks to fierce radiation from nearby stars 'exciting' its atoms and molecules, causing them to emit light with specific wavelengths and colours.

Extrasolar planet
A planet beyond our solar system, orbiting another star.

Galaxy
An independent system containing a large number of stars (many millions or billions) in orbit around a common centre – often a supermassive black hole.

Giant
A star that has evolved away from its place on the 'main sequence' of stellar evolution, increasing in luminosity but at the same time expanding and cooling.

Light
A form of electromagnetic radiation visible to the human eye. Light with different wavelengths and energies is detected as different colours.

Light year
A common unit of measurement in astronomy, equivalent to the distance travelled by electromagnetic radiation (such as light) in a year – equivalent to 9.5 million million km or 5.9 trillion miles.

Magnitude
(Technically 'apparent magnitude'). A measure of the perceived brightness of a star in visible light – affected by its luminosity, surface temperature, distance from Earth and the presence of intervening material to absorb its light.

Main sequence	The longest period in the life of any star, during which it shines by nuclear fusion of hydrogen in its core to form helium.
Moon	A natural satellite of a planet. Moons may form from the same cloud of collapsing material as their parent planet, or be captured by its gravity at a later stage. The Moon (with a capital 'M') refers specifically to Earth's satellite.
Multiple star	A system of two or more stars in orbit around one another. Members of multiple star systems account for the majority of stars in the Milky Way galaxy.
Nebula	A cloud of gas and dust floating in space, often the scattered remains of previous generations of stars, from which new stars can form. Nebulae are usually classed by their appearance as dark, emission, or reflection nebulae.
Neutron star	The collapsed core of a massive star that had destroyed its outer layers in a supernova explosion. A neutron star is typically the size of a city, with the mass of several Suns. They are often also pulsars.
Nova	An enormous explosion on the surface of a white dwarf star that has pulled matter away from its companion star in a binary system.
Nuclear fusion	A process that takes place only at extreme temperatures and pressures, in which the nuclei (central cores) of light atoms are forced together to form heavier ones, releasing energy in the process. Nuclear fusion is the process that powers the stars.
Oort cloud	A spherical halo filled with billions of icy dormant comets, surrounding the solar system between half a light year and one light year from the Sun.
Open cluster	A loosely bound cluster of dozens to hundreds of stars, born at the same time from a star-forming nebula, and usually dominated by brilliant but short-lived blue stars.
Planet	An object in an independent orbit around the Sun (or another star) that has sufficient mass to pull itself into a spherical shape, and whose gravitational influence has cleared its orbit of other objects. See also Dwarf planet.
Planetary nebula	The outer layers of a giant star with relatively low mass, shrugged off in glowing shells as the dying star exhausts its fuel supplies and ceases nuclear fusion.
Pulsar	A neutron star with a powerful magnetic field that channels its radiation into two beams. As the star rotates, the beams sweep across the sky like a cosmic lighthouse.
Red shift	A stretching of the wavelength of radiation (sending it towards the red end of the spectrum) caused when the radiation's source is moving rapidly away from the observer.

Reflection nebula	A nebula that shines by reflecting the light from nearby stars.
Rocky planet	A planet like Earth, which is largely composed of rock and metals with high melting points, typically found close to the Sun.
Satellite	Any object that orbits around another one – natural satellites that either formed alongside their parent planet or were captured into orbit after formation are known as moons.
Solar system	The region around the Sun, and everything that orbits within its gravitational grasp.
Spectral line	A bright ('emission') or dark ('absorption') line in the spectrum of light from an object, created where energy is released or absorbed at a specific wavelength. Patterns of spectral lines can be used to trace the presence of specific atoms or molecules.
Spectrum	A rainbow-like pattern made by splitting the light from a star or other object through a prism or similar device, so that the light is spread out according to its wavelength and colour.
Star	A massive ball of light gases (chiefly hydrogen at first) whose core is hot and dense enough to allow nuclear fusion reactions to take place. Radiation forcing its way out from the core balances the inward pull of gravity and prevents the star collapsing completely.
Sun	The star at the centre of our solar system.
Sunlike star	A star with broadly similar properties to the Sun, which is likely to follow a roughly similar evolutionary path.
Supergiant	One of the brightest stars in the Universe, extremely massive and typically hundreds of thousands of times as luminous as the Sun.
Supernova	A powerful stellar explosion that may briefly outshine an entire galaxy. Some supernovae mark the death of the most massive stars, while others are caused when a white dwarf in a nova system gains enough mass to collapse further into a neutron star.
Supernova remnant	The wreckage of a supernova explosion, consisting of an expanding cloud of shredded gas enriched with heavy elements, with a superdense object – a neutron star or black hole – at its centre.
Universe	The entirety of creation, encompassing every galaxy in existence and all of space and time. The Universe was born in the Big Bang about 13.6 billion years ago, and is still expanding from this initial explosion. However, the speed of light places a limit on the size of the observable Universe.
White dwarf	The collapsed core of a relatively low-mass star that has ceased nuclear fusion and shrugged off its outer layers in a planetary nebula. The more massive a white dwarf is, the smaller and denser it gets, until above 1.44 solar masses, it collapses into a superdense neutron star.

Index

INDEX

INDEX